A Transformation Gap?

A Transformation Gap?

AMERICAN INNOVATIONS
AND EUROPEAN MILITARY CHANGE

Edited by

Terry Terriff, Frans Osinga, and Theo Farrell

Stanford Security Studies

An Imprint of Stanford University Press

Stanford, California 2010

Stanford University Press
Stanford, California

© 2010 by the Board of Trustees of the Leland Stanford Junior University.
All rights reserved.

Printed in the United States of America on acid-free, archival-quality paper

Library of Congress Cataloging-in-Publication Data

A transformation gap? : American innovations and European military change / edited by Terry Terriff, Frans Osinga, and Theo Farrell.
     p. cm.
Includes bibliographical references and index.
ISBN 978-0-8047-6377-6 (cloth : alk. paper) —
ISBN 978-0-8047-6378-3 (pbk. : alk. paper)
   1. Europe—Armed Forces—Technological innovations.   2. Europe—Armed Forces—Reorganization.   3. Military art and science—Technological innovations—Europe.   4. North Atlantic Treaty Organization—Europe.   5. Europe—Military relations—United States.   6. United States—Military relations—Europe.   I. Terriff, Terry.   II. Osinga, Frans P. B.   III. Farrell, Theo.

  UA646.T728 2010
  355'.07094—dc22   2009040654

Typeset at Stanford University Press in 10/14 Minion

Special discounts for bulk quantities of Stanford Security Studies are available to corporations, professional associations, and other organizations. For details and discount information, contact the special sales department of Stanford University Press.
Tel: (650) 736-1783, Fax: (650) 736-1784

CONTENTS

Acknowledgments

As befits a book on a transatlantic theme, this project started as a series of conversations between the editors in California and London. We received much help along the way in developing the project and producing this volume. First, we must thank the authors for sharing their ideas over several workshops, and responding with such good cheer to editorial guidance. Everybody pulled together on the project, and a real team spirit developed. It was a delight to experience. We thank King's College London and Allied Command Transformation (ACT) for sponsoring project conferences in 2007 and 2008 respectively. At ACT, we especially thank NATO's Supreme Allied Commander Transformation, Gen. James Mattis, and the Deputy Chief of Staff for Transformation, Lt. Gen. James Soligan, for supporting our project. We are grateful to the participants at these workshops for their feedback on the draft chapters. We also gratefully acknowledge feedback on papers presented at project panels at the International Studies Association and International Security Studies Section annual conventions. We wish to gratefully acknowledge financial support for this project from the Economic and Social Research Council (Grant RES-228-25-0063). Terry Terriff would further like to express his appreciation to the Arthur J. Child Foundation for their financial support for his work on this project, and Theo Farrell similarly would like to acknowlege the financial support he received to work on this project from a research fellowship funded under the UK Research Council's Global Uncertainties Programme (ESRC Grant RES-071-27-0069). Finally, we wish to thank the team at Stanford University Press, Jessica Walsh, John Feneron, and especially our editor, Geoffrey Burn, whose commitment and enthusiasm for the project matched our own. And that's saying something.

Contributors

THE EDITORS

Dr. Terry Terriff is the Arthur J. Child Chair of American Security Policy in the Department of Political Science and is also a Senior Research Fellow of the Centre for Military and Strategic Studies at the University of Calgary. He has published widely on NATO, transatlantic security, and change in military organizations, most recently on military change in the U.S. Marine Corps and U.S. Army.

Dr. Frans Osinga is a Colonel in the Royal Netherlands Air Force. He has recently completed a tour at Allied Command Transformation, NATO, and is now Assistant Professor at the War Studies Department of the Royal Netherlands Military Academy, Breda.

Dr. Theo Farrell is Professor of War in the Modern World in the Department of War Studies at King's College London. He was Associate Editor of *Security Studies* (2005–9) and is the Security Studies Editor of the ISA Compendium Project.

THE AUTHORS

Dr. Antonio Marquina is Chair in International Security and Cooperation and Director of UNISCI at Complutense University, Madrid. He is also President of the Asia-Europe, Human Security Network.

Dr. Tim Bird is Lecturer in the Defence Studies Department of King's College London. From September 2008 to July 2009 he was on secondment to the UK

Ministry of Defence as part of the writing team producing new UK military stabilization doctrine.

Dr. Heiko Borchert directs business and political consultancies in Switzerland and Austria, is a member of the advisory board of IPA Network International Public Affairs, Berlin, and is affiliated with The Hague Centre for Strategic Studies. He has worked with military and national security organizations in Switzerland, Austria, and Germany.

Gustavo Díaz has an MA in Intelligence and Security Studies from Salford University. He is Academic Director of Interligare and is currently a researcher at UNISCI, Complutense University.

Dr. Rob de Wijk is Director of The Hague Centre for Strategic Studies. He is also Professor in the field of international relations at the University of Leiden. He was previously Head of the Defence Concepts Division at The Netherlands Ministry of Defence.

Dr. Sten Rynning is Professor of Political Science at the University of Southern Denmark. He is a former Fullbright and NATO Research fellow.

Olaf Osica (Ph.D.) is a senior research fellow at the Natolin European Centre in Warsaw and a senior lecturer in the Department of International Relations at the Collegium Civitas in Warsaw.

A Transformation Gap?

Military Transformation in NATO: A Framework for Analysis

Theo Farrell and Terry Terriff

Across the North Atlantic Treaty Organization (NATO), member states are undergoing military transformation. For new member states from the former Eastern bloc, there has been transformation with the introduction of democratically controlled military organizations.[1] Some existing member states have also been transforming their militaries to phase out conscription and move toward all-volunteer professional forces.[2] On top of these de facto transformations in military professionalism across Europe, NATO undertook a commitment at the 2002 Prague Summit, to transform its capabilities for, and approach to, military operations, and to lead that effort a specific NATO command was created, Allied Command Transformation (ACT).

Notwithstanding the establishment of ACT, military transformation has been a US-led process centered on the exploitation of new information technologies in combination with new concepts for "networked organizations" and "effects-based operations" (EBO). European states have simply been unable to match the level of US investment in new military technologies, and so for some time critics have warned of a growing "transformation gap" between the United States and the European allies.[3] In recent years, this process of developing transformational technologies and concepts for war has been reoriented toward tackling counterinsurgency (COIN) and stability operations. Here, the experience (especially from colonial times) of states such as Britain and France gives some European militaries a possible transformation advantage over the big war orientated US military.

This study assesses the *extent* and *trajectory* of military transformation across a range of European NATO member states. It considers cases of the major European military powers (Britain, France, and Germany), smaller Western militaries (Spain and The Netherlands), and one new member state (Poland). This study offers a more nuanced picture of the much touted transatlantic "transformation gap." It shows the enormous variation among the European allies on the extent of transformation, suggesting that there may be both technological and conceptual gaps *within* Europe.[4] It describes how a complex and contingent mix of international and local drivers is operating to push forward military transformation in each country. And, accordingly, it provides insight into the variable trajectories of military transformation among NATO member states.

In this chapter, we introduce the common analytical framework for the book. Essentially, this centers on breaking transformation down into three discrete elements -- namely, network-enablement, effects based operations, and expeditionary warfare. The extent and trajectory of military transformation in each case-study is assessed in terms of these respective technological, doctrinal, and organizational innovations. A secondary goal of this volume is to assess the process of military transformation in each country. This involves an enormously complex interaction among international, national, and organizational factors in each case. Nonetheless, three discrete scholarly literatures considered in this chapter, on military innovation, norm diffusion, and alliance theory, respectively, do provide pointers for analyzing the processes of military transformation.

WHAT IS MILITARY TRANSFORMATION?

At the turn of the twenty-first century, the United States officially embarked on a process of military transformation. Chapter 2 of this volume discusses the rise of military transformation in detail, and thus we provide only a brief introduction in this section to orientate the reader. In the 2001 Quadrennial Defense Review (QDR), the US Department of Defense (DoD) declared transformation to be "at the heart" of its "new strategic approach."[5] Military transformation is rooted in the US-led Revolution in Military Affairs (RMA) of the 1990s, when it became apparent that advances in information technology (IT), as harnessed by a resource-rich US military, offered the potential to revolutionize the conduct of warfare, in much the same way that mechanical transport, metal steam-powered ships, and manned flight all revolutionized warfare. The spectacular American victory in the 1991 Gulf War also seemed to suggest that a US-led RMA was underway and that US forces were leaping forward in military capability.[6] Indeed, while a number of NATO states provided air and naval forces to support the US-led coalition, only Britain and France provided significant ground contingents. Moreover, the French 6th Light Division was allocated to the far left flank, well away from the main coalition land offensive, because, unlike the British 1st Armoured Division, it was not deemed to be up to major combat operations.[7] In large part, then, the Gulf War appeared to be a success for the RMA, and militaries around the world began to look closely at the new US military model based on the exploitation of IT. However, no state could hope to match the level of US investment in IT-enabled military capabilities.[8] So while the rhetoric of the US RMA spread rapidly to other militaries, actual emulation of the new US military model, insofar as it occurred, did so selectively and on a surface level only.[9]

By the end of the 1990s, the term "military transformation" had begun to replace that of RMA in describing the program of change in the US military. This shift in terminology highlighted that this process of revolutionary military change involved as much new thinking as new technology. Whereas RMA was focused mostly on IT, transformation was equally focused on the new operational concepts and organizational forms that would enable the US military truly to revolutionize the conduct of warfare. Hence, in 2003 the US Secretary of Defense, Donald Rumsfeld, defined transformation in this way:

> [A] process that shapes the changing nature of military competition and cooperation through new combinations of concepts, capabilities, people and organizations that exploit our nation's advantages and protect against our asymmetric vulnerabilities to sustain our strategic position which helps underpin peace and stability in the world.[10]

Thus, for Rumsfeld, transformation could conceivably involve all manner of things, including a return to cavalry—with US special forces operating on horseback in Afghanistan calling in precision air strikes.[11] Crucial to transformation was new agility in doctrinal thinking, organizational form, and operational approach. In short, transformation involved nothing less that a paradigmatic shift in the US approach to warfare. Accordingly, Rumsfeld called on the military to foster "a culture of creativity and prudent risk taking."[12]

Just as the 1991 Gulf War focused worldwide military attention on the future of conventional warfare, so ongoing coalition operations following the 2001–2 Afghan War and 2003 Iraq War have focused Western military minds on the return of irregular warfare. This, in turn, has had an impact upon the transformation debate, resulting in divergent perspectives within the US on the content and direction of military transformation. Broadly speaking, it is possible to discern two strands—one focused on conventional warfare, the other on COIN and stability operations. The first strand, more commonly called "Force Transformation," has been the most prominent. As noted above, it has involved innovating concepts, organizations, and technologies for major combat operations. At the same time, in November 2005, the Undersecretary of Defense (policy) issued a directive requiring the (DoD) and military service to give stability operations "priority comparable to combat operations."[13] In truth, it is hard to imagine the army, air force, and navy giving equal weight to forms of operations (such as stability operations) that essentially threaten the purpose of traditional prestige weapon platforms: heavy armor, strategic bombers, and aircraft carriers. Nonetheless, operational demands and experience in Iraq and Afghanistan, and new US military doctrine on small wars and COIN, reinforce the genuine interest in the US military in what some have called "Second Generation Transformation."[14]

Both strands of US military transformation are producing innovations—both in terms of technology and ideas—that are of interest to the European allies. As suggested above, the Europeans have been concerned for a number of years about the "transformation gap," and this is leading them to consider how they need to transform their own militaries in order to be able to continue to fight alongside the US military in conventional wars. At the same time, European militaries (especially those with colonial and peacekeeping experience) will be naturally drawn to COIN and stability operations. And here there has been some two-way transatlantic traffic of military ideas. In 2004–5, the US Army was keen to learn from the European (especially British) experience of COIN. Since 2006, the US Army and Marine Corps have developed new concepts, doctrine, and organizational capabilities, building on lessons learned from Iraq and Afghanistan.[15] The new joint US Army/US Marine Corps doctrine on COIN (FM 3-24) has greatly influenced doctrinal development and the conduct of operations by European militaries.[16]

THE ANALYTICAL FRAMEWORK

This study adopts a focused, structured comparison approach to examining military transformation in six NATO member states.[17] Accordingly, we focus on specific aspects of each case and analysis is structured by a common set of research themes concerning the extent, process, and trajectories of military transformation. Our case selection provides for analysis of military transformation in a range of European member states. Britain, France, and Germany, as the major powers in Europe, have similar potential to resource military transformation. At the same time, these three cases offer variation in terms of political and military ties to the United States, recent operational experience, and national military culture. We have also selected two additional existing member states—one northern (The Netherlands) and one southern (Spain), and one new member state (Poland). In all three of these additional cases the states are (or have until recently been) sympathetic to US grand strategy, and the militaries have recent operational experience with US forces.

At the heart of both transformation strands are two innovations, network-centric warfare (what the Europeans call network-enabled capability) and effects-based operations. There is also a major organizational change—namely, the shift from predeployed forces for territorial defense of Europe to forces configured for expeditionary warfare. Assessment of the extent of transformation in each case concentrates on these three elements of transformation. Empirically, this involves looking at change in doctrine, training, and force structure, as well as at future force development and systems acquisition plans in each country.

Network-Enabled Capability (NEC)

This innovation originates in the US concept for network-centric warfare (NCW). This is the notion that a "system of systems," connecting sensors, information processing centers, and shooters operating as one network across the whole of the battlespace, will replace platform-centric warfare conducted by large, self-contained military units. The Europeans prefer the term "network-enabled capability" (NEC), with "enabled" intentionally replacing "centric"; this indicates the more modest expectations (itself reflecting the more modest resources) of the Europeans in terms of the transformative effect of networking on military organization.

Effects-Based Operations

This concept originated in US Air Force thinking in the early 1990s about the future of striking air power. In its original formulation, EBO was about reconsidering how operational effects could be most efficiently produced through air strikes. Thus, whereas previously the focus had been on destroying targets, under EBO, it may be sufficient or indeed preferable to disable targets. In the late 1990s, the meaning and scope of EBO changed considerably. It was adopted more widely by all the US services and increasingly by US allies, and was redefined more broadly in terms of focusing military operations on the campaign objective—that is, the strategic effect. Arguably, this broader definition is rather meaningless (because military operations should always concentrate on campaign objectives), though it does have the virtue of focusing military attention on how the conduct of the campaign (particular in terms of the level of destruction) contributes or retracts from campaign objectives. The language of "effects-based" operations is now widely used in NATO even if there is some differing on precise terms and meanings. Indeed, NATO's Military Committee formally adopted the concept of the effects-based approach to operations (EBAO) in 2006.[18] Like EBO, EBAO focuses attention on the strategic effects of operations but also places priority on the integration of the various instruments of the alliance and coordination with other international organizations.

Expeditionary Warfare

This element involves the most change—in organizational structure and capabilities—for the land forces. Expeditionary warfare has always been a core mission for the US Marine Corps. But now it has become a core mission for the US Army. This is most evident in the US Army's program for restructuring from ten divisions to forty-three brigade combat teams (BCTs). There will also be a sizable number of support brigades, but essentially BCTs will be self-sufficient combat formations. A typical army division had around 15,000 troops, whereas a BCT has between 3,000 and 4,000. Thus this new modular force structure promises to give the US Army a more agile force structure, and one that is better suited to expeditionary warfare. It so happens that a number of European militaries undertook similar restructuring programs in the mid to late 1990s, to do away with unsustainable legacy force structures and to introduce new self-sustainable and deployable units; hence Britain began to fold army regiments into battalions, while France broke up its divisions into maneuver regiments. The development of expeditionary forces needs also to be viewed in light of the innovative Combined Joint Task Force concept (CJTF), adopted by NATO in 1994. Introduced to NATO by the US military, CJTFs were the early version of the current drive to create modular expeditionary forces.[19] So we would expect to find considerable evidence of this aspect of military transformation in many of the case studies.

PROCESSES OF MILITARY TRANSFORMATION

Military transformation involves both external processes of military emulation and internal processes of military innovation. To be sure, a central concern of this volume is with the question of whether, and to what degree, European military transformation has been influenced by new military concepts and ideas from the United States. Since the end of the Cold War, the United States has been the unquestioned dominant military power. Moreover, as noted already, the 1991 Gulf War was an apt demonstration of US military prowess. Hence, the US military has provided an exemplar, or template, for European states seeking to transform their militaries for the information age. Driving this European interest in transformation is also a traditional concern with sustaining their alliance with the United States. European states recognize that, should they fail to keep up with the Americans, they will become militarily irrelevant.

In this section we consider three scholarly literatures that are relevant to our analysis of the process of military transformation in individual European states. First is the literature on military innovation, which suggests the factors that shape domestic processes of military transformation. Second is the literature on norm diffusion, which explores the transnational processes whereby US ideas and innovations spread to European militaries. Third is the literature on alliance politics, which points to traditional intra-alliance concerns with interoperability and burden sharing.

Military Innovation

Since military transformation involves a number of innovations, the military innovation literature is of obvious relevance to case studies in this volume.[20] This literature broadly suggests three main factors that shape the trajectory of military innovations: threat, civil-military relations, and military culture. Related literatures on bureaucratic politics and weapons procurement suggest that entrenched institutional interests can also be a barrier to military innovation.

Threat is an obvious spur to military innovation. Indeed, the dominant theoretical approach to International Relations theory (that is, Realism) stresses how alert states are to threatening changes in the international system.[21] Threatened states may respond by forming alliances, but they may also internally mobilize military resources, and that, in turn, may require military innovation. Indeed, the threatening development may itself be military innovation by an opponent (a new weapon or way of war) that demands an innovative response.[22] States will be particularly sensitive to external military threats in the shadow of war. Moreover, defeat or major setback in war can itself push militaries to innovate.[23] Even militaries that are victorious in war but at very great cost may seek to innovate in order to make future victory less costly. Thus, a British general staff horrified by the human toll of World War I embraced new technologies in the interwar period in order that the next battles might be won "with the maximum of machinery and minimum of manpower."[24]

The military innovation literature also highlights the role of internal factors. With regard to civil-military relations, the key issues are military responsiveness to civilian policy, and the ability of civilian policymakers to effect military innovation. The literature recognizes that civilians are often more ready to contemplate major military change in response to new policy priorities or strategic challenges, precisely because they do not have an organizational stake in existing military practices, equipment, and structures. However, opinion is divided on the question of whether civilian intervention is effective, or even required, in military innovation. One school argues that because militaries are slow to change, civilians must intervene to force them to innovate (especially in peacetime), and that such innovation is most likely to succeed when civilian policymakers team up with maverick military experts.[25] An alternative school argues that militaries must and do innovate by choice. Innovation requires visionary military leaders, who enjoy legitimacy within their organizations by virtue of their formal position, to lead campaigns to effect major change within their organizations. Externally imposed change will fail, it is argued, because civilians lack the knowledge, and mavericks the authority, to effect real change.[26]

The third shaping factor is military culture. Military culture comprises those identities, norms, and values that have been internalized by a military organization, and that frame the way the organization views the world, and its role and functions in it.[27] Military culture is embodied in (and reproduced through) military training, regulations, routines, and practice. Given that military practices, technologies, and structures usually reflect cultural bias, as a general rule we may expect military culture to act as a brake on innovation. That is the point of the military-led innovation school discussed above: since innovation involves changing those things that militaries take for granted, you need a leader with authority to champion what is, in effect, cultural change. Even with such a champion, innovation is unlikely to succeed if it challenges the core identity of the organization.[28] Thus of two innovations attempted by two different commandants of the US Marine Corps (USMC)—the adoption of maneuver warfare as the overarching war-fighting concept for the USMC under General Alfred Gray (1987–91), and the creation of a "culture of constant innovation" under General Charles Krulak (1995–99)—the former succeeded because it was consistent with the "warrior identity" of the Corps, while the latter failed because it was not.[29] For this reason, innovation that goes against organizational identity usually requires some external shock to military culture, such as defeat in war, in order to jolt the military into a fundamental rethink of its purpose and core business.[30]

Related to the military innovation literature is scholarship on bureaucratic politics and on weapons acquisition that underlines the institutional and political barriers to innovation. The large literature on bureaucratic politics,[31] like the work on military culture, points to characteristics of military organizations that produce obstacles to truly innovative military change. Often these vested interests are located in dominant war-fighting communities within organizations (for example, promoters of armor in the army, bomber and fighter pilots in the air force, surface warfare officers in the navy). Military innovation almost always challenges vested organizational interests: new ideas threaten familiar operational practices, and new technologies threaten legacy systems. Crucially, military innovation may threaten the traditional flow of resources.[32] Related to bureaucratic politics are broader entrenched interests that drive military procurement. Studies of weapons procurement highlight military biases toward developing high-performance and high-cost equipment, and the means the military have to advance those biases: institutional longevity, knowledge asymmetry of defense matters, and manipulation of the program process (through things like sunk costs and gold-plating).[33] These studies also point to domestic politics (favoring home-produced systems), national industrial capacity, and industry-government links as factors that influence program choice and outcome.[34]

Norm Diffusion

From the European perspective, military transformation is a transnational process involving an emerging norm of military organization and operations that originates from the United States. As such, the literature on international norm diffusion is relevant here. Much of this literature is concerned with the spread of norms of state sovereignty, good governance, and human rights.[35] More recently, some scholars have begun to look at norm diffusion in the military sphere.[36] The norm diffusion literature suggests external processes of military change that are shaped by a number of considerations.

The first consideration is motive. Essentially, with norm diffusion we are not talking about military innovation so much as military *emulation*. The military in question innovates by emulating another—usually we are talking about developing states emulating more developed ones. The norm diffusion literature suggests two motives for emulation. The first is success: militaries will emulate those that have been successful in battle. This is a rationalist account of emulation which suggests that strategic imperatives provide the driving motive. In this sense, the dramatic US victory in the 1991 Gulf War provided the perfect "poster child" for the US model of an IT-intensive military. The other motive is legitimacy: militaries emulate the strongest states, even if such military models are inappropriate for poorer or less threatened states, because strength confers legitimacy. Hence even the militaries of the smallest states, which had little need and could ill afford an IT-intensive military, were emulating the US military model in the 1990s. This is a cultural account of emulation which suggests that legitimacy imperatives provide the driving motive.[37]

The second consideration is norm strength, which may be measured in terms of how well defined the norm is, how long-established it is, and how widely accepted it is.[38] Arguably, transformation has involved the diffusion of two norms, one involving process and the other content. The process norm is one of "constant innovation": transformation aims to produce innovative militaries. Transformation is also about, as we have suggested, a specific mix of technological and doctrinal innovations and organizational changes. In this sense, we may say that there is a fairly well defined "content norm" that specifies network-enabled, effects-orientated, and expeditionary organizations and operations. Both norms are widely accepted by Western militaries. However, our study shows, the process norm is accepted more in rhetoric than in reality. And there is enormous variation in actual interpretation, at a national level, of the content norm. Both norms are also only a few years old, and so neither are well established.

The third consideration is the transmission structure for new international norms. The norm diffusion literature points to the role of institutional structures—legal regimes, international organizations, epistemic communities, and policy networks—in promoting, transmitting, and sustaining the norms in question.[39] Obviously, for this study NATO is the most important institutional structure and, in particular, Allied Command Transformation. Bilateral military relations may also be a significant form of transmission structure. This is especially true for Britain, with its close military ties to the United States, but other potential significant bilateral relations include those between France and Germany. For the European member states, the European Union may also be a transmission structure for new military norms.

The fourth consideration in norm diffusion is the end-point of the process—that is, norm internalization. Crucial to transnational norms taking root in national contexts is the degree of match or mismatch between the new norms and existing military culture. Also important is the degree of match with strategic culture—that is, beliefs shared and practiced by national policy communities about when and how to use force.[40] Strategic culture is broader than national military ways of war. It covers the fundamental beliefs of a national polity about the place and purpose of force in that country's foreign policy.[41] This raises the question of how military transformation, with its model of a high-tech military postured for expeditionary operations, fits the various strategic cultures of European states.[42] Broadly speaking, where there is a cultural clash, mere rhetorical observation if not outright rejection of the new military norms is likely. However, driving strategic imperatives to introduce new ways of war may override pre-existing military cultural biases. We may also expect to see some degree of adaptation of new transnational norms to suit local conditions in the process of internalization.[43] Such local conditions include military culture, political imperatives, economic considerations, and bureaucratic interests.

Alliance Theory

Alliance theory mostly concerns the conditions that lead states to operate in alliances; these include foreign and domestic security threats,[44] as well as shared ideology and norms.[45] However, this literature does suggest certain conditions that also might affect the processes of military transformation in NATO.

The first is the impact of variable transformation on the institutional efficiency of NATO. Alliances produce efficiency in the generation of military power when states are able to mass force through operational collaboration, and realize economies of scale through collaboration on military procurement and logistics.[46] The emerging "transformation gap"—both between the US and Europe, and possibly between the major and minor European states—thus threatens alliance efficiency by reducing the capacity for multilateral operations and common procurement and logistics.

The second condition is the politics of burden-sharing within alliances. At the heart of alliances is an alliance security dilemma, whereby states fear abandonment but also at the same time fear entrapment.[47] This dilemma is most acute when it comes to crises and war. But at a more minor level it also operates in terms of force development in the problem of burden-sharing. Here there is also a dilemma, between the fear of contributing too much and being taken for a ride (similar to the fear of entrapment), and the temptation to contribute too little and take others for a ride. Restraining this temptation is the fear of abandonment: that if one overtly free-rides, disillusioned alliance partners may end collaboration.

The third condition is the two-level game involving domestic and alliance politics.[48] For the governments of developing states, alliances (especially with more powerful allies) promise to bring resources—social and material—that may be used to bolster political stability.[49] For governments facing internal challenges to their authority, "internal balancing" concerns will outweigh the "external balancing" concerns posited by realist theory.[50] These domestic political concerns may be especially evident, therefore, in the case of new member states of NATO, who may use the legitimacy and security apparatus of NATO to secure themselves politically. In this sense, participating in an overall alliance program of military transformation increases the legitimacy connection between the new member and NATO. In developed states, domestic politics also plays in a role in terms of producing pressures that may run counter to the direction of alliance policy. Alliance theory also points to the strategic use of domestic politics by policymakers for leverage in debates with alliance partners.[51]

In closing, we note that external threat is an important condition that operates in tandem with others listed above. It can give added impetus to the drive for institutional efficiency. It also affects the burden-sharing problem by making the fear of abandonment more acute, and thereby reducing the attraction of free-riding. However, the alliance theory literature does suggest that far more important to the long-term functional effectiveness of an alliance is commonality of interests and ideology. The historical record suggests that alliances formed out of shared fear and nothing else don't work well and rarely survive.[52]

In each of the chapters that follow, case study experts analyze the extent and trajectory of military transformation in their respective national militaries, paying particular attention to the development of EBAO, NEC, and expeditionary warfare capabilities. Each chapter also considers the process of transformation in that country, looking at the interaction of international forces and local conditions, and the mix of strategic, political, and cultural influences that explain why and how European militaries are transforming. The literatures discussed above point to key themes that emerge in many of these chapters. One theme is the importance of military culture, civil-military relations, and resource considerations in shaping European military transformation. Another theme is the diffusion of US military innovations and how those innovations are received, adapted, and adopted by European militaries. A final theme is the imperative for European members of NATO to respond to US burden-sharing concerns, and the "legitimacy benefits" of military transformation for new members of NATO. A concluding chapter returns to these themes in exploring the cross-country patterns and national variations in European military transformation.

2 The Rise of Military Transformation

Frans Osinga

Armed forces cannot just change these days; instead, they must transform. Instigated by Secretary of Defense Donald Rumsfeld, US armed forces have embarked on a deliberate program to improve the agility, lethality, and expeditionary capabilities of the US military. European militaries too must now transform. After the Prague Summit of November 2002, "Transformation" became institutionalized within NATO, and operationalized with the creation in 2003 of Allied Command Transformation (ACT) in Norfolk, Virginia. NATO's Transformation program is an ambitious institutional effort—and an imperative in the eyes of US political leaders—to get European nations to converge in their defense policies in a very particular direction.

The first deputy commander of ACT, the British Admiral Ian Forbes, emphasized the strong connection with the US Transformation program when he stated in 2003 that "[a] transformational process akin to that which has been taking place in the United States is essential to modernize the Alliance's capabilities and ensure that they stay consistent with US military thinking and development."[1] Within many European armed forces, Transformation (with a capital T) is now a formally accepted idea, objective, program, topic of academic debate, pretext for reprioritization of investments, and the main reason for the existence of several new defense organizations.

As Transformation captures recent developments in US military technology and ideas and the declared aspiration of European nations, Transformation—the US and NATO version of it—offers us a useful template for approaching the issue of contemporary military innovation and change in Europe. Moreover, NATO is an obvious and important avenue of infusion of US military ideas and technology. For many European nations, NATO is one of the prime inter-national institutions through which they collectively respond to international security issues. Many smaller nations in particular, and those nations that have recently joined NATO, also look to NATO to inform and sometimes justify their national defense policies. NATO furthermore acts as an "agent for change," a role it has deliberately and explicitly endorsed with the creation of ACT.

Transformation is therefore a useful tool—a conceptual lens—for approach-ing the question of whether European nations are in fact explicitly adopting US military ideas and technology, and to what extent. Subsequently, tracing the history and trajectory of US transformation can inform us about the expected trajectory of European armed forces if they indeed want to make good on their stated intentions. If European nations use Transformation as an idea to inform their future, the history of US Transformation will show us the mold, and the future of European militaries lies in no small measure in the history of US transformation. However, Transformation has also proven to be a fluid idea, changing somewhat over time, and US Transformation is similar to, but still different from, NATO Military Transformation.

This chapter charts the rise of Military Transformation in both the US and European contexts. It begins by using official US documents to develop a per-spective on the meaning, key concepts, and implications of US Military Trans-formation. Second, the chapter will show how those key concepts, which are at the heart of Transformation, emerged from developments in the 1990s and resulted in a specific way of warfare. The final part of the chapter discusses the European context in which the idea landed in 2002, and shows how Transfor-mation in NATO gained a slightly different, or rather, additional meaning.

THE US POLICY PERSPECTIVE

The term "defense Transformation" came into common use in the late 1990s and has been defined in various ways. One dominant perspective stresses the nature of the process of change. Prompted by significant changes in technol-ogy or the emergence of new and different international security challenges, this perspective describes transformation in terms of comprehensive, discon-tinuous, and possibly disruptive changes in military technologies, concepts of operations (that is, approaches to war-fighting), and organization, in contrast to incremental or evolutionary change that marks normal defense moderniza-tion. In 2003 the US Department of Defense (DoD) in one document defined transformation in this way:

a process that shapes the changing nature of military competition and cooperation through new combinations of concepts, capabilities, people and organizations that exploit our nation's advantages and protect against our asymmetric vulnerabilities to sustain our strategic position, which helps underpin peace and stability in the world. [...] It does not have an end point. Transformation anticipates and creates the future and deals with the co-evolution of concepts, processes, organizations, and technology. Profound change in any one of these areas necessitates change in all.[2]

In addition, the process of Transformation would touch noncombat aspects such as training, personnel management, logistics, and worldwide basing arrangements, and affect DoD business policies, practices, and procedures, geared toward achieving efficiencies and reducing costs as well as the time between developing and fielding new weapon technologies.

In the wake of Operation *Enduring Freedom* in Afghanistan, Transformation also gained operational substance and bureaucratic traction. In a widely quoted article in *Foreign Affairs,* Donald Rumsfeld saw in the Afghanistan operation both a validation of the idea of Transformation and the future trajectory of it.[3] In this perspective, Transformation concerns improving mobility, agility, and lethality. Key to this process was the further development and implementation of technologies and organizational adjustments to effect network-centric warfare (NCW), a concept that emerged in 1998 (as will be described below). Second, ground forces would in the future need to operate increasingly like special operations forces (SOF). Meanwhile, the improved joint interoperability achieved through NCW capabilities would foster precision-strike operations. Acting hand in glove with these tenets was the third element of Transformation: making US military forces more expeditionary.

These notions found their way into DoD plans. In 2003, Rumsfeld developed six critical operational goals that would focus transformation efforts: (1) protecting critical bases and defeating chemical, biological, radiological, and nuclear weapons; (2) projecting and sustaining forces in anti-access environments; (3) denying enemy sanctuary; (4) leveraging information technology; (5) ensuring information systems and conducting information operations; and (6) enhancing space capabilities. Through the Office of Force Transformation, created by Rumsfeld and headed by the conceptual father of the NCW concept and Transformation proponent, Vice-Admiral Arthur Cebrowski, the US DoD promulgated a number of documents that laid out the plans for defense transformation.[4] They called for shifting the US military away from a reliance on

TABLE 2.1

The Changing Security Environment

From	To
A peacetime tempo	A wartime sense of urgency
A time of reasonable predictability	An era of surprise and uncertainty
Single-focused threats	Multiple, complex challenges
Nation-state threats	Decentralized network threats from nonstate enemies
Conducting war against nations	Conducting war in countries we are not at war with
"One size fits all" deterrence	Tailored deterrence for rogue powers, terrorist networks and near-term competitors
Responding after a crisis starts	Preventive actions so problems do not become crises.

massed forces, sheer quantity of firepower, military services operating in isolation from one another, and attrition-style warfare, and toward a greater reliance on joint (that is, integrated multiservice) operations, NCW, Effects-Based Operations (EBO), speed and agility, and precision application of firepower. These changes constituted a shift from an industrial-age approach to war to an information-age approach.[5] But whereas Transformation initially was technology driven with an eye on improving combat *and* cost effectiveness, from 2003 onward Transformation was seen as an integral part of the administration's wider policy response to the new security environment post 9/11, an environment that the 2005 *Quadrennial Defense Review*, submitted to Congress on 6 February 2006, described in terms of shifts (see Table 2.1).[6]

The same document subsequently states that these shifts in the environment require changes—Transformation—in the structure and modes of operations of US armed forces (see Table 2.2).

The military services and DoD agencies subsequently developed Transformation plans (or road maps). The Army's Transformation plan centered on reorganizing the Army into modular, brigade-size forces called Units of Action (UAs) that can be deployed to distant operating areas more easily and can be more easily tailored to meet the needs of each contingency. Key elements of the Air Force's Transformation plan included reorganizing the service to make it more expeditionary, and exploiting new technologies and operational concepts to improve dramatically its ability to deploy and sustain forces rapidly, to dominate air and space, and to rapidly identify and precisely attack targets on a global basis. Finally, Naval Transformation centered on operating in littoral

TABLE 2.2

Consequences for Defence Planning

From	To
A focus on kinetics	A focus on effects
Twentieth-century processes	Twenty-first-century integrated approaches
Static defense, garrison forces	Mobile, expeditionary operations
Under-resourced, standby forces	Fully equipped combat ready units
A battle-ready force (peace)	Battle-hardened forces (war)
Large institutional forces (tail)	More powerful operational capabilities (teeth)
Major conventional combat operations	Multiple irregular, asymmetric operations
Separate military service concepts of operation	Joint and combined operations
Forces that need to deconflict	Integrated, interdependent forces
Emphasis on ships, tanks and aircraft	Focus on information, knowledge, and timely, actionable intelligence
Massing forces	Massing effects
Set-piece maneuver and mass	Agility and precision
Single service acquisition systems	Joint portfolio management.
Service and agency intelligence	Truly Joint Information Operations Centers
Vertical structures and processes (stovepipes)	More transparent, horizontal integration
Moving the user to the data	Moving data to the user
Predetermined force packages	Tailored, flexible forces
Department of Defense solutions	Interagency approaches

(that is, near-shore) waters, new-design ships requiring much smaller crews, directly launching and supporting expeditionary operations ashore from sea bases, more flexible naval formations, and more flexible ship-deployment methods. All transformation plans stressed greater jointness and implementing NCW.[7]

THE MOTHER OF TRANSFORMATION: DESERT STORM

The logic of the argument of the Transformation initiative follows and flows directly from the rapid developments in military technology of the 1990s, which, some argued, constituted nothing less than a Revolution in Military Affairs (RMA). Transformation in that sense can be seen as the culmination of fifteen years of rapid military changes within the US armed forces and can also be equated with the emergence of the so called New American Way of War. And arguably this can be traced back to Operation *Desert Storm* (the Gulf War,

1991). While the RMA and ideas embedded in network-centric warfare can be traced back to Vietnam, the teachings of John Boyd and the AirLand Battle concept, *Desert Storm* can be considered the spark plug of much of the debate on the RMA, and subsequent efforts to make it a reality.[8] Rightly or wrongly, *Desert Storm* was taken to represent a new age of warfare, and high-ranking US politicians such as Dick Cheney and William Perry saw a revolutionary advance in military capabilities in the Gulf War.[9] *Desert Storm* was a watershed because the way it unfolded surprised even US military experts. The thirty-nine days of massive yet precise air attacks preceding the four-day ground campaign was a break with the common and expected pattern of operations.[10] Another factor was the "CNN effect." For the first time the entire world could witness the effectiveness of modern Western military systems on television just hours after the actual attacks, and sometimes even in real time when cruise missiles were recorded buzzing through the streets of Baghdad. A new image was created.

Two icons stood out in this image.[11] The first was the demonstrated advance achieved in precision in detection, identification, and attack capabilities. *Desert Storm* heralded the age of precision warfare (remarkable considering that only 5 percent of all ordnance dropped was actually precision guided). Attacks employing precision-guided munitions (PGMs) proved thirteen times as effective as nonprecision attacks.[12] Stealth was the second icon, and for most people, even military experts, a novelty in its effectiveness and strategic value. It has been labeled, with some justification, revolutionary. With a radar reflection surface similar to that of a golf ball, F-117 stealth fighter aircraft could operate almost unseen deep in enemy territory from the first moment of the war, sometimes attacking two targets per mission in the Baghdad area, which sported the highest density of air defense systems in the world.

The new dominance of offense over defense in air warfare through the use of stealth, stand-off weapons, electronic warfare, and drones offered a sanctuary that could be exploited for various purposes. And even nonstealth aircraft, if equipped with precision munitions and precision information, could steer clear of even advanced air defense systems by flying at high altitude while maintaining accuracy of attacks. *Desert Storm* demonstrated that advanced air power capabilities offer the option to open a flank in the third dimension. Intense day and night air attacks with PGMs offered the possibility of relatively quick success against old-fashioned armed forces relying on massed mechanized ground combat. In addition, relentless PGM attacks on trucks and bridges was effectively used to halt the flow of supplies to the Iraqi frontline and the movement of units within the Kuwaiti theater. This drastically shortened the time required and the risk involved for ground units to complete the coalition victory.

Moreover, precision, stand-off, and stealth capabilities offered new possibilities for strategic attacks against multiple target-categories of a nation-state (for example, military units, leadership, and critical infrastructure). Even if targets were in the vicinity of civilian objects, it was possible to attack these near-simultaneously in order to rapidly degrade the functioning of the entire "enemy system." Instead of the traditional model, in which a country's defeat required first a decisive victory over the armed forces, *Desert Storm* heralded the "inside-out" model. The overwhelming air power capabilities offered the potential to strike at the heart of a country (the regime) from the first moment of a campaign and cripple the strategic command capabilities before attacking fielded forces.[13]

Several lessons were drawn from *Desert Storm*. The demonstrated new air power capabilities promised "overwatch" over a crisis area or during posthostilities, such as was practiced later over Iraq during Operations *Southern* and *Northern Watch*. It appeared that diplomacy could now be bolstered by a credible force able to strike accurately, at short notice if necessary, without undue risk of casualties. It also promised the ability "to seize and maintain the initiative, to dominate the course of hostilities, to deny the adversary the ability to force an alteration in US strategy and to foreclose its pursuit of strategic alternatives, and the capacity to defeat adversary forces in the field."[14]

Desert Storm also hinted at implications for force structures and joint doctrine. As Eliot Cohen observed, platforms would become less important, while the quality of what they carry—sensors, munitions, and electronics of all kinds—would become critical. In addition, the quality and speed of the command process itself was becoming a war-winning element. *Desert Storm* was heralded as the first information war. The flow of secure, rich, relevant, and timely information, and the denial of it to the opponent, was increasingly becoming a decisive front. Moreover, the age of mass warfare was regarded as drawing to an end. It appeared that "everything that moves can be seen and everything that can be seen can in principle be hit."[15] Massing of ground troops and armored units was becoming more and more dangerous and outdated. The industrial-age warfare model that had existed since World War I now seemed to be surpassed by another—yet to be defined—model.

Another implication was that military campaigns could be designed differently in the future. The enormous effects of air attacks on ground units suggested that, in future, ground operations should begin only after optimal conditions have been created by air attacks so as to minimize risk. In the Kuwaiti theater, coalition air attacks managed to destroy sometimes more than 50 percent of Iraqi armor and artillery equipment, and Iraqi ground troops surrendered by the thousands after being pounded by B-52 strikes or leaflets threatening such attacks. The correlation of this trend was the suggestion that control of territory could no longer be equated or ensured with physical presence of ground troops. Nonlinear warfare could be envisioned in which small teams of ground troops would operate deep within enemy territory in close coordination with air power, replacing linear warfare, defined by long, closed frontlines of army units advancing slowly. It indicated that joint doctrine, too, had to be amended, to reflect the insight that the ground phase of a military campaign could start much later and would be a function of the effectiveness of the air campaign.

The experience of *Desert Storm* also held another promise for the future. It suggested that military operations need not necessarily entail massive civilian casualties and that "collateral damage" to civilian infrastructure could be contained. In addition, the risk for coalition troops was lower than expected. Approximately 148 coalition military personnel died in combat, a regrettable but also unprecedentedly low number considering the scale of the operation and the pessimistic prewar estimates of 10,000 coalition casualties.[16] Indeed, the Gulf War departed from others in its speed, scope, and relative "cheapness" in terms of casualties. Keaney and Cohen concluded that "the ingredients for a transformation of war may well have become visible in the Gulf War, but if a revolution is to occur, someone will have to make it."[17]

THE EMERGENCE OF NETWORK-CENTRIC WARFARE

And that is exactly what the US military and its industrial suppliers set out to do. Ever since the Gulf War, experiments and concept development aimed to fully exploit the technological advances that *Desert Storm* foreshadowed.[18] Prime objectives were to make the battlefield more transparent, to achieve "information dominance" and create situational awareness at all command levels, to disseminate target information in a timely manner to those who needed it, and to adjust command and control doctrine accordingly. The objective was to shorten the "sensor-to-shooter" time, and to improve responsiveness. In short, the US military aimed to improve military effectiveness on three different axes: lethality, visibility, and agility.[19]

Three technological streams were instrumental in this effort. First, information technology: the rapid increase in computing power and transmission capabilities of modern communication systems offered the opportunity to analyze, disseminate, and access unprecedented quantities of information in ever shortening time. Efforts were directed at fusing data-streams originating from different units, services, nonmilitary governmental organizations such as the CIA, and from different sensor platforms (satellites, Early Warning [EW] aircraft, Unmanned Aerial Vehicles [UAVs], forward air controllers) into coherent "pictures" in command centers offering greater situational awareness. Experiments explored ways to organize the flow of information, to eradicate organizational barriers to information access, and to define the appropriate level of (de)centralization of command in light of the increasing availability of information at lower command levels and the consequences of operating over vaster distances.

The second and closely related technological stream was the development in surveillance and sensor capabilities. Detecting, observing, and tracking objects of military concern during all weather, day and night, on a routine basis became increasingly feasible, also for nonspecialized air and ground combat systems. Tanks, armored personnel carriers, and individual soldiers gained night vision equipment, in addition to GPS location devices, data links, and mobile computer displays, all improving their situational awareness. Night precision air attacks and all-weather/beyond-visual-range air combat operations used to be the preserve of specialized aircraft. New radar systems, onboard infrared sensors, and improved navigation equipment brought these within reach of aircraft such as the F-16 and A-10, originally designed as simple, lightweight, day fighter aircraft.

The 1990s also saw the (albeit reluctant) rise of a new generation of UAVs that operate at medium and high altitude, mainly at the operational level of war, in contrast to previous systems that were merely for tactical artillery spotting. These new vehicles could perform dull, dirty, and dangerous reconnaissance missions over enemy territory for twenty-four to forty-eight hours nonstop. Initially equipped for photoreconnaissance, the emphasis gradually shifted to multispectral sensor suites and to realizing the ideal of real-time "streaming" of video through various data-links to other aircraft and command centers. Operating at 65,000 feet, the large Global Hawk UAV can survey 40,000 square miles and focus on 1,900 spot targets in twenty-four hours by day or night under all weather conditions, with a resolution of 30 centimeters from a distance of 100 kilometers.[20] These developments translated into an improved ability to spot, identify, and track potential or actual targets no matter what their speed and, if necessary, provide the information through data-links in real time to command centers and weapon platforms.

The third stream of technological developments concerned airframes and air armament, or rather, the ability to hit targets precisely and quickly.[21] Stealth technology was further refined. Not only was stealth accorded with immense operational value, it was also deemed a measure of indispensable efficiency improvement. *Desert Storm* indicated that, whereas a typical nonstealth attack package required thirty-eight aircraft to enable eight of those to deliver bombs on three targets, only twenty F-117s were required to simultaneously attack thirty-seven targets successfully, in the face of an intense air defense threat. Precision-guided munitions were improved, with average miss distances reduced to three to ten feet by the end of the decade and stand-off range constantly increasing. The cost of PGMs went down dramatically, too. While one cruise missile costs more than $1 million, the latest generation of JDAM (Joint Direct Attack Munition) weapons with GPS guidance "sells" for $20,000. These developments improved the efficiency of attacks. With state-of-the-art systems, one strike aircraft, be it a B-2 bomber or an updated F-16, could strike several targets on one mission from a stand-off range outside the threat envelope of surface-to-air missile (SAM) systems. Instead of large-scale destruction of targets, precision information and precision weapons allow for the achievement of measured effects.

The digital battlefield was thus in the making. Information was becoming the driving factor in warfare;[22] indeed, two RAND analysts predicted in 2003 that "Cyberwar Is Coming."[23] They noted that sea changes were occurring in how information is collected, stored, processed, communicated, and presented, and in how organizations are designed to take advantage of increased amounts of information.[24] Thus, they claimed, success in warfare was no longer primarily a function of who puts the most capital, labor, and technology onto the battlefield, but of who has the best information about the battlefield. What distinguishes the victors is their grasp of information, not only from the mundane standpoint of being able to find the enemy while keeping it in the dark, but also in doctrinal and organizational terms.[25] Organizations should adapt their structures and processes toward flexible, network-like models of organization. The information revolution favored the growth of networks by making it possible for diverse, dispersed actors to communicate, consult, coordinate, and operate together across greater distances, and on the basis of more and better information than ever before.[26]

By 1997, US Defense Secretary William Cohen had asserted, "The information revolution is creating a Revolution in Military Affairs that will fundamentally change the way US forces fight."[27] Joint Vision 2010 condensed information age warfare tenets and the US defense aspirations as follows:

> By 2010, we should be able to change how we conduct the most intense joint
> operations. Instead of relying on massed forces and sequential operations, we
> will achieve massed effects in other ways. Information superiority and advances
> in technology will enable us to achieve the desired effects through the tailored
> application of joint combat power. Higher lethality weapons will allow us to
> conduct attacks currently that formerly required massed assets applied in a se-
> quential manner. With precision targeting and longer range systems, command-
> ers can achieve the necessary destruction or suppression of enemy forces with
> fewer systems, thereby reducing the need for time consuming and risky massing
> of people and equipment. Improved command and control, based on fused, all-
> source real-time intelligence will reduce the need to assemble maneuver for-
> mations days and hours in advance of attacks. Providing improved targeting
> information directly to the most effective weapon system will potentially reduce
> the traditional force requirements at the point of main effort. All of this suggests
> that we will be increasingly able to accomplish the effects of mass—the neces-
> sary concentration of combat power at the decisive time and place—with less
> need to mass forces physically than in the past.[28]

The vision was of a small, rapidly deployable, highly accurate, stealthy, highly
lethal, extremely well skilled, and less costly force.[29] It capitalized on various
service level experiments such as the US Navy *Cooperative Engagement Concept* (CEC), which built upon the "system of systems" concept. According to its
author, Admiral Owens, this was the emerging mode of US warfare. If US systems could be better integrated, that could potentially "lift the fog of war" in a
battle-space of 200 by 200 nautical miles.[30] According to the CEC, the US Navy
would link sensors of all ships in a battle group together with airborne and
space-based assets to provide an increase in situational awareness and engagement capability in each ship without increasing the sensor suite. The US Army
experimented with digitization and reorganization of its brigades in its Force
XXI initiative. The USAF created Air Expeditionary Forces and funded the development of new Combined Air Operation Centers with a heavy emphasis on
modern IT support. Furthermore, the USAF successfully attempted to data-
link sensors, communication systems, and weapon delivery platforms with the
aim to improve the situational awareness of air operation centers as well as

projecting real-time target data and images into the cockpits of bombers. This would allow for "flex targeting" whereby aircraft could be retasked in flight or be provided with data of emerging targets while orbiting in a patrol area.

From 1998 onward these ideas found a home in an overarching concept, Network-Centric Warfare, which subsequently became the formal guiding framework for shaping the future of the US armed forces.[31] Summarizing the NCW advantages, Paul Murdock lists the following:[32]

- NCW could permit a geographically dispersed force to operate as a system—in effect, as a "dispersed mass." Such a force, though its elements might be spread over a large area, should be able to concentrate precision weapons rapidly upon targets hundreds of miles away.
- Its units may be able to mass fires not only with decisive effect but without the need to maneuver—without having to get closer to targets, avoid geographical constraints, or achieve some positional advantage.
- NCW offers the flexibility, operational reach, and battlespace awareness needed to operate on the strategic, operational, and tactical levels at once. Combat would no longer have to proceed in the traditional step-by-step, or serial, manner. Combat would instead be multidimensional and comprehensively joint.

Indeed, with NCW the Pentagon embraced the belief and the tenets of the RMA thesis. NCW, according to its advocates, is the "emerging theory of war in the information age," "a paradigm shift," "the military embodiment of Information Age concepts and technologies."[33]

TOWARD EFFECTS-BASED OPERATIONS

What and how to strike was also subject of debate. Inspired by the promise of new technologies of stealth and precision, John Warden, one of the principal designers of the air campaign, and one of his assistants, Dave Deptula, elaborated upon the strategic utility of intense precise coercive air attacks on "leadership" targets, and the paralyzing theater-level employment of air power as demonstrated in *Desert Storm*. The concept of *Parallel Warfare* was debated, which called for simultaneous attacks on the enemy's key systems, or centers of gravity, so as to paralyze them.[34] Attacking various interrelated nodes could create a ripple effect across an enemy system, direct or indirect, be it in the physical, cognitive, or moral domain. They argued that, instead of focusing on armed forces exclusively, with modern air power other options were available as well, such as regime targeting and precise infrastructural disruption that could have direct, strategically significant effects. Parallel warfare also held the promise that several smaller-scale physical and nonphysical attacks—with less destructive power—conducted simultaneously could achieve disproportionate strategic outcomes.

This academic exercise met real-world needs when US commanders encountered operational and strategic problems in the Balkan conflict. This inspired a debate among US academics and military planners on the best strategy for coercing an opponent (whereas in Europe much doctrinal discussion centered on the principles of peacekeeping).[35] This debate concerned the questions of what and how to target, when, and for how long. Although the parameters of defeating a mechanized army in a traditional high-intensity war were quite well understood, in particular after *Desert Storm*, the opposite was true of the dynamics of coercing unwilling leaders such as the Serbian leader Milosevic with conventional force in a limited conflict. Various "coercive mechanisms" were discerned and advocated, such as *decapitation* and *incapacitation* (paralyzing the country or its military apparatus by eliminating command nodes or disrupting command processes), *punishment* (increasing the cost of achieving a strategic aim), *denial* (eliminating the means to carry out the strategy, thus decreasing the chances of success), *second-order change* (threatening a higher order interest than the values originally at stake), or *hybrid strategies* (combining these). The intensity of attacks was also a topic of debate, with one doctrinal school advocating "decisive force," massively and continuously applied for maximum political and military shock, while others favored a gradually increasing intensification so as to provide room for political maneuver.

Eventually, these various schools of thought on targeting were conceptually tied into an overarching concept called Effects-Based Operations, which became part of US joint targeting doctrine in 2001. It recognized that US forces must be able to produce a variety of desired military and political effects, not merely destruction. Tailored to the type of conflict and the specific political objectives, an EBO-based strategy aims to produce distinctive and desired effects through the application of appropriate movement, supply, attack, defense, and maneuvers. Effects-Based Operations focus on functional, systemic, and psychological effects well beyond the immediate physical results of a tactical or operational undertaking. This requires detailed and up-to-date knowledge of the behavior of various subsystems of the opponent. Consequently, the ability to plan and conduct Effects-Based Operations is predicated on a task force operating along the principles of the Network-Centric Warfare concept.[36] When Transformation was launched by Rumsfeld, he was thus building upon a stream of technological and doctrinal developments spanning a decade.

MIDCOURSE VALIDATION AND VINDICATION

Two consecutive operations seemed to validate those developments, and thus Transformation, in the eyes of the US administration as well as scores of analysts. First, in Afghanistan in 2001, the odds had been distinctly against the favored low-risk, high-tech type of warfare of the West. The US was confronted with an enemy trained in guerilla fighting in mountainous terrain, with an impressive track record against the former Soviet Union and domestic rivals, with no significant infrastructure offering strategic coercive leverage, and within a region nonsupportive of US military action. It was neither obvious nor predetermined that the US would come out victorious from Operation *Enduring Freedom*, and with such relatively low costs in terms of destruction and losses. Only 300 to 500 special forces actually operated within Afghan territory, uniting, empowering, and fighting alongside local opposition factions totaling no more than 15,000 men. This combination of US troops and proxy forces managed to evict a force of 60,000 Taliban fighters and the regime.[37] This required a relatively limited operation of 100 combat sorties a day, amounting to 38,000 sorties flown. Outside Afghanistan, a US/UK force of approximately 60,000 personnel supported this operation, dispersed over 267 bases, in 30 locations in 15 countries. The US lost 30 personnel.[38] And again, the use of PGMs increased, this time up to 60 percent, indicating that PGMs had become the norm. As the commander, General Franks, asserted, this was by far the greatest application of precision munitions in the history of his country.

Importantly, the operation reflected the merits of the NCW concept discussed above.[39] The integration of ground-air communications was unprecedented and represented a revolutionary operational concept.[40] Combat aircraft, dispersed air bases, command centers, and special forces were glued together by a network of sensors and communication systems. In the opening phase of the air campaign, fixed targets (roads, bridges, and command facilities) had been struck, limiting the Taliban's ability to communicate, move, disperse, reconverge, and attack unobserved and unhindered. Afterward, attention shifted to so-called emerging targets such as small Taliban troop contingents.[41] JSTARS (Joint Surveillance and Target Attack Radar System), UAVs, and special forces acted as eyes, spotting pop-up targets and relaying time-sensitive, up-to-date, accurate target information to shooter platforms inbound or already circling in the vicinity. Considering the ways of the opponent this C4ISR capability was indispensable,[42] as Franks later asserted.[43] It offered a stunning reaction capability, with response times sometimes down to several minutes, and averaging only twenty minutes. Rumsfeld thus saw in *Enduring Freedom* a revolution similar to the blitzkrieg concept.[44] In both, old and new technology was employed in innovative ways. US Defense Under Secretary Paul Wolfowitz proudly agreed and told the US Congress that "the capabilities demonstrated in Afghanistan show how far we have come in the 10 years since the Persian Gulf War."[45] For the Office of Force Transformation, it redefined "The American Way of War."[46]

Compared with Afghanistan, Iraq provided an even better test ground for the new concepts. The planning for Operation *Iraqi Freedom* was dominated by a clash between new and old thinking. Actually it was a clash between the advocates of modern and postmodern warfare.[47] Especially the US Army presented heavy options with big, mechanized divisions. Most force packages presented to Rumsfeld were rejected as "too big." The secretary put his trust in NCW with precision bombing, a small, fast-moving ground attack force, and heavy reliance on special operations forces and air power.[48]

The Americans confirmed that the combination of innovative concepts and power projection with high-tech forces for advanced expeditionary warfare was able to achieve objectives with astonishingly low numbers of friendly casualties and modest collateral damage. Air-ground surveillance systems, unmanned aircraft, and SOF located conventional Iraqi forces while a continuous stream of fighter aircraft delivered ordnance on the accurate target locations they were provided. As a result of superior intelligence and the number of available offensive air assets over the area at any one time, it took approximately twelve minutes to destroy a confirmed target; in some cases it was five minutes after detection. In the west and north of Iraq large numbers of SOF teams operated as part of a closely integrated team with airborne sensors, command nodes, and offensive aircraft to detect and neutralize potential launches of surface-to-surface missile such as the Scud, and to restrict Iraqi freedom of movement on the ground.

Networking of forces contributed to the tempo. The combination of intensive air strikes with the highly mobile ground force continued day and night, and small, fast-moving forces defeated larger forces. There were only 125,000 personnel in Iraq with only three divisions forming the "spear" of the attack, while Iraqi forces numbered 400,000 including some 100,000 well-trained and

-equipped Republican Guard troops. In a single week, the coalition destroyed 1,000 tanks and reduced the Republican Guard by 50 percent. Equally important was the effect on Iraqi military morale, which effectively collapsed as coalition air strikes caused increasing attrition to Iraq forces. Precision strikes combined with psychological operations caused Iraqi units, including one entire armored division, to dissolve.[49] Vice Admiral Cebrowski observed tighter integration between land and air operations. Logistical support was equally impressive. The United States not only managed to fight halfway around the globe; it could also move ammunition, fuel, and water to maneuver units deep within the theater of operations. A new model of warfare thus emerged; its merits were seen as validated, and a revolution seemed to be confirmed—or at least such a narrative was plausible.[50]

TRANSFORMATION HITS NATO

The events of 9/11 served as a catalyst in the sense that for the US the time had come to press forward forcefully within NATO the need for military change commensurate with the changes in the geopolitical environment. In various NATO summits and meetings from December 2001 till the one in Prague in November 2002, Rumsfeld made it clear that, if NATO were to remain a relevant organization, it needed not only to embrace new missions but also to make good on the initiatives to improve Europe's military capabilities so as to be able to execute those missions. Over Kosovo, US forces accounted for 60 percent of all sorties, dropped 80 percent of all expended ordnance, and provided 70 percent of all support sorties and 90 percent of all suppression of enemy air defenses and electronic warfare sorties—not to mention the fact that without US support NATO would have lacked effective command facilities.[51] In response, at the 1999 Washington summit, NATO launched the Defence Capabilities Initiative (DCI), which listed fifty-eight shortfalls, divided into areas of deployability, sustainability and logistics, effective engagement, survivability of forces and infrastructure, command and control, and information systems. Six areas of high priority were identified, involving strategic lift, air-to-air-refueling, suppression of enemy air defenses, support jamming, precision guided munitions, and secure communications.

The policy initiative, however, had not gained much traction in practice. Although many blamed the costs of modernization, the heart of the problem was not money but policy reorientation and force restructuring.[52] Through the 1990s most European armed forces had not changed their orientation on territorial defense. Europe still had 1.5 million people in arms, and in excess of 10,000 tanks. On the other hand, only 10 to 15 percent of those troops were actually deployable.[53] If NATO could barely manage Operation *Allied Force*, a small-scale operation of limited complexity, what did that mean for more demanding operations?

The Prague Capabilities Commitment of 2002 unsurprisingly read like a repetition of DCI, but now it was more focused (some would say also more limited) and included specific, ambitious, quantified goals and timelines. This also reflected the widening of NATO's geopolitical envelope, a trend not quite unrelated to the shifting US security political agenda under the Bush administration. From 2003 onward NATO's scope became defined by a range of up to 10,000 kilometers from Brussels, and since the Istanbul Summit of 2004 the alliance has made overtures toward, for instance, Australia and Japan, while continuing gradually to expand its network of members and partners east and southward. De Hoop Scheffer stated in Munich early in 2006, "We have broadened our strategic horizon far beyond Europe, and tackling terrorism, engaging it at the source, is now a main mission."[54] Threats now include failed states, radical ideologies, unresolved conflicts, and criminal networks trafficking in people, drugs, and weapons, while energy security is also being considered.[55] During the Riga Summit of November 2006 this trend was continued. Again it was noted that the alliance needed the ability to respond to challenges "from wherever they may come." NATO needs to be able to face weapons of mass destruction (WMD) and other asymmetric threats, and attacks that may originate from outside the Euro-Atlantic area.[56]

Missions now include humanitarian relief operations, and in the wake of the US experience in Iraq and that of NATO in Afghanistan, counterinsurgency, and stabilization and reconstruction. At the 2004 Istanbul Summit, defense against terrorism had already been added as a key priority. As NATO Secretary-General Robertson had noted in 2003, this new security environment required "not a sumo wrestler" but "a fencer—light, fast, able to adjust quickly and strike precisely."[57] In 2004 his successor, de Hoop Scheffer, warned his audience in a similar vein that NATO needed "forces that are slimmer, tougher, and faster; forces that can reach further, and stay in the field longer but still can punch hard."[58]

NATO Transformation has thus gained a specific political content. When NATO documents state that NATO Transformation is about the future of the alliance, this may be understood as justification and hope, but also as a threat. Indeed, NATO Transformation is in no small measure a renewed and downscaled attempt to solve the problems that became glaringly evident during Operation *Allied Force*, and to get beyond the period of "dynamic stagnation" in terms of European military modernization and capabilities improvement.[59]

NATO MILITARY TRANSFORMATION

The scope and importance of capability improvement and modernization under the banner of NATO Transformation was expressed squarely by the first commander of ACT, US Admiral Edmund Giambastiani (a Rumsfeld protégé),[60] when he told an audience that it involved bringing changes to doctrine, organization, capabilities, training, and logistics, and would be significant, both culturally and intellectually. The prize, he noted, would be improved interoperability, fundamentally joint, network-centric, distributed forces capable of rapid decision superiority and massed effects across the battlespace, critical to continuing alliance relevance.[61] While broad, overused, and underdefined, recent NATO publications are also quite specific concerning the meaning of Transformation, taking their cues from the New American Way of War. Inspired by the much discussed high-level document *Concepts for Allied Future Joint Operation*, the ACT pamphlet *Understanding NATO Military Transformation*, published in 2005, for instance, notes that Transformation encompasses reorientation and reorganization of force structures, redefines the way combat power is generated and employed, and leads to new ways to approach and conduct military operations, thus addressing the capability gap between the US and Europe and hence between actual and required capabilities vis-à-vis the new security environment.[62]

Overall, NATO Transformation calls for expeditionary capabilities, as NATO will most likely operate from austere bases that are potentially under threat, at strategic distances, in a variety of environments, including urban jungles. Several specific benchmarks are listed, such as scalable command, control and communications (C3) assets that can span large distances, precision, speed, agility, and the ability to disperse rapidly and concentrate force and forces. NATO operations must also be conducted in a way that minimizes unintended damage as well as the risk to our own forces. All this calls for lighter ground units than ever before, reconfigurable and mobile, equipped with more precise and effective firepower, all the while ensuring a smaller logistic "footprint" than before so as to minimize the threat to vital supply lines. That in turn implies sizable air capabilities to provide in-theater transport, surveillance, interdiction and intervention options, as well as offensive and force protection capabilities.

It requires European air forces to broaden their range of capabilities so as to go beyond their Cold War tactical focus, to include strategic lift, precision attack, and stealth capabilities. Navies will have to include the littoral as the most likely operating environment and invest in sea lift.

Such a concept, moreover, presumes modern 'command, control, communications, computers, military intelligence, and surveillance, target acquisition, and reconnaissance' (C4ISTAR) and networking capabilities to enhance situational awareness, timely operational planning, and decision-making, and to improve "the links between commanders, sensors and weapons" as the Comprehensive Political Guidance, agreed upon in Riga, noted with priority.[63] NATO needs information superiority, a notion captured in the *NATO Network-Enabled Capability (NNEC)* concept. NNEC differs from the US NCW idea in the sense that European armed forces are reluctant to put networks at the core of war-fighting, for both financial and operational reasons, but it acknowledges that increased investments in, and exploitation of, latest generation C4ISTAR technology are much needed, overdue, and operationally very promising. One cannot escape the impression that many of the NNEC publications have been unashamedly informed by US NCW publications as published by Rumsfeld's two in-house think tanks, Office of Force Transformation (OFT) or Office of the Secretary of Defense/Command and Control Research Program (OSD/CCRP).

ACT, meanwhile, has tried to gain acceptance within NATO of the concept of the Effects Based Approach to Operations *(EBAO)*, which includes the narrow military EBO concept. It is based on the idea of coherence and interdependence, and the realization that peace, security, and development are more connected than ever.[64] Security can be achieved only when threats are dealt with in coordination with other international and nongovernmental organizations, the Riga Summit communiqué noted. EBAO recognizes that, apart from the military instrument, there are three other instruments that need to be coordinated in concert: political, economic, and civil. The political instrument refers to the use of political and diplomatic power to influence an actor or to create conditions that are advantageous to the alliance. It involves efforts within and among the various regional and international organizations and actors. The use of the economic instrument refers generally to financial incentives or disincentives. This instrument is most likely to be exercised not by NATO but by nations or international organizations. The civil instrument refers to areas such as judiciary, constabulary, education, public information, and civilian administration and support infrastructure, which can lead to access to medical care, food, power, water, and administrative capacities of nations and nongovernmental organizations. Recognizing that a preponderance of nonmilitary capabilities are not at the direct disposal of NATO, the EBAO concept encourages NATO to work to achieve coherence among the actions of various agencies and organizations toward the achievement of effects that are beneficial to the ends of NATO's particular operation. In Oslo on March 2006, de Hoop Scheffer noted that intense interagency cooperation was already a reality on the ground, whether in the Balkans, Afghanistan, Iraq, or Darfur, stressing, however, that these ad hoc methods of cooperation need to become structured relationships at the institutional level—to be able to coordinate strategically, not just tactically.[65] NATO has thus elevated EBO out of its military environment and expanded it. While EBAO may thereby have become "just" another bureaucratic acronym that really means nothing more than a call for a capacity for proper (grand) strategy making, it again points at the influence of US ideas.

Several instruments are, in principle, at NATO's disposal for effecting change. First, it can strive to improve coherence among national defense investment plans through the revamped "Defense Requirements Review" process. Second, through concept development and experimentation, and multinational exercises, through the constant updating of the standards concerning training, tactics, techniques, and procedures, and through developing new doctrine and concepts, it hopes to get nations to converge in their defense policy orientation and investment priorities, and to foster closer cooperation and interoperability. Finally, the NATO Response Force (NRF) was created, to which nations must contribute with units that would of necessity be forced to become interoperable. Thus the NRF is considered both a rapidly deployable force as well as a catalyst for Transformation.

CONCLUSION

US "Transformation" emerged from a series of interlinked conceptual and technological developments over the past twenty years, showing the interplay of experience, debate, technological developments, and policy development. The conceptual and technological developments have been funneled into NATO, and are captured in the term NATO "Transformation." In NATO, Transformation refers to the following objectives:

- Closing the so-called capability gap between the US and Europe;
- Catching up on the RMA of the 1990s—that is, modernizing European military technology and accelerating the process of technological, doctrinal, and organizational innovation;
- Improving the expeditionary capabilities of European armed forces;
- Adopting—more of less—the New American Way of War, including ideas such as Effects-Based Operations and Network-Centric Warfare;
- Improving and thereby ascertaining interoperability with US armed forces.[66]

Indeed, if NATO's documents of the past few years can serve as a guide, it is quite clear what European defense policies should aim for, and the ideas captured in those documents strongly suggest that the future of Europe's militaries can in no small measure be found in the narrative of US military experience over the past fifteen years. Conceptually it is but a small step from US Transformation to NATO Transformation. Whether the US-driven push through NATO to insert American ideas and technologies into European armed forces will succeed in practice is another matter. As noted in the first chapter, security and defense policy is heavily shaped by strategic as well as domestic political and cultural factors, and a myriad of intervening domestic variables may quite well lead nations onto other pathways. To what extent, why, and how European nations actually adopt US ideas and technology will become evident in the next chapters.

Innovating within Cost and Cultural Constraints: The British Approach to Military Transformation

Theo Farrell and Tim Bird

In a world that is perceived to present an increasing array of operational challenges, the British armed forces have embraced a vision of transformation that comprises all three elements identified in this volume: network-enabled forces, the effects-based approach to operations, and expeditionary warfare. The Defence Secretary, Geoff Hoon, provided a clear statement of the direction of British military transformation in his foreword to the 2003 Defence White Paper (DWP):

> Our focus is now on delivering flexible forces able to configure to generate the right capability in a less predictable and more complex operational environment. This will require us to move away from simplistic platform-centric planning, to a fully "network-enabled capability" able to exploit effects-based planning and operations, using forces which are truly adaptable, capable of even greater precision and rapidly deployable.[1]

Of course, civilian policymakers and military leaders may disagree on the direction of military transformation. Indeed, as noted in Chapter 1, there is considerable debate in the military innovation literature over the role of civilian intervention in managing military change.[2] In the British case, however, there is consistency between civilian policy and service plans for force development. Hence, the Royal Navy's future operational concept for its Versatile Maritime Force (VMF) notes, "Effects Based approach, facilitated by Network Enabled Capability (NEC) will drive all aspects of VMF development."[3] Equally high on the list of the Chief of the Air Staff's strategic priorities out to 2020 is the development of "agile, adaptable and capable expeditionary air power" able to "deliver effects across the battlespace by developing and exploiting the UK's network enabled capability."[4] Finally, the Future Land Operational Concept (FLOC), which guides the development of the British Army out to 2020, centers on the core concepts of agile forces, the effects-based approach to operations (EBAO), and NEC; "directed logistics" is the fourth core concept. Also of note is that the types of operations imagined in FLOC—military intervention, power projection, and peace operations—all involve expeditionary warfare.[5]

Insofar as this vision of transformation is American in origin, and it largely is, British emulation of it is hardly surprising. States, given their resources, invariably emulate the military innovations of the leading world power, especially one that has enjoyed victory in war. The spectacular coalition victory in the Gulf War dramatically validated (or appeared to validate) the emerging US model of an information-technology (IT) enhanced military.[6] But in addition to these international pressures for military emulation is British political and military proximity to the United States. Britain, of all the European allies, enjoys the closest strategic and military ties with the United States; from Britain's perspective, this constitutes an enduring "special relationship." Indeed, one should not forget that Britain and the United States formed an exceptionally close alliance during World War II. In addition to uniquely close military ties continued throughout the Cold War, there has been recent experience of joint campaigning (in the 1991 Gulf War, the 1999 Kosovo War, the 2001–2 Afghanistan War, and the 2003 Iraq War). Thus the British were almost bound to embrace the US vision of military transformation.

Naturally, military transformation has also been driven by another critical external factor—namely, strategic change and budget cuts following the end of the Cold War. In the early 1990s, Britain found itself with a heavy army stuck in Germany, a navy geared up for antisubmarine warfare, and an air force equipped to repulse Soviet bombers. A major mismatch soon developed between these Cold War legacy force structures and the post–Cold War missions centered on power projection in the Gulf and peace operations in Europe and Africa. Added to this new strategic context was the political and public expectation of a post–Cold War peace dividend. In response to the changed strategic environment, Conservative governments conducted two defense reviews—*Options for Change* in 1990–91, and *Front Line First* in 1994—both of which focused on cutting defense costs.[7] Far more influential, however, on the current trajectory of British military transformation was the Strategic Defence Review (SDR) conducted by the incoming New Labour government in 1997–98. The SDR noted that defense expenditure had been cut by 25 percent and the armed forces by nearly a third in 1990.[8] Hence SDR was deliberately set up not as a cost-cutting exercise but a radical rethink of the future role and shape of Britain's armed forces. The SDR established the policy provenience for what was to become British military transformation: more agile, joint, expeditionary, effects-oriented, and technology-enabled forces. As we shall see, these lines of post–Cold War military change have been reinforced in successive defense policies, most recently and significantly the DWP, as well as military plans and programs.

This chapter explores, in turn, the development of the three elements of British military transformation: NEC, EBAO, and expeditionary warfare. In each we find the *external drivers* of transformation outlined above: US military innovations or post–Cold War strategic change, or both. But also in evidence for each element are *shaping factors* that are internal to Britain, especially resource constraints and military culture. We conclude the chapter with some general observations concerning the trajectory of British military transformation.

NETWORK-ENABLED CAPABILITY

The Ministry of Defence (MOD) acknowledges that the 2003 DWP clearly placed NEC "at the heart of our transformation." NEC is defined as "the coherent integration of sensors, decision-makers, weapon systems and support capabilities to achieve the desired effect."[9] NEC is, not surprisingly, also prominent in the UK Joint High Level Operational Concept (Joint HLOC), specifically in the core concepts of inform and command.[10] Crucially, the Joint HLOC emphasizes the significance of NEC as an "essential element" of EBAO and expeditionary capability.[11] For the purpose of analysis, these three elements of transformation are separated out in this chapter but in reality they are interlinked.

NEC occupies a similar place in the future operational concepts of all three services. The Royal Navy expects NEC to deliver "a common high-quality information domain, shared situation awareness, and contribution and real-time access to the Joint Operational Picture (JOP)."[12] The Royal Air Force's future operational concept emphasizes the central importance of air- and space-based ISTAR[13] systems to NEC, and specifically in terms of generating JOPs. It also boldly declares that "the full potential of EBO can only be unlocked by NEC."[14] Of course, navy and air force enthusiasm for NEC is entirely to be expected. NEC is premised on new technology, and navies and air forces tend to be more technocentric than armies.[15] More to the point, perhaps, networked-enabled operations originated in sea and air operations during World War II, in the

US and Japanese carrier warfare in the Pacific and the Royal Air Force's radar-directed campaign in the Battle of Britain.[16] However, the British Army has also embraced NEC, which it sees as a "key enabler" of operational capability. Indeed, the FLOC identifies NEC as the "core concept that unifies the other [concepts]" of agile forces, EBAO, and directed logistics.[17]

A driving concern for the British, in developing network capability, is interoperability with the Americans. Key to this is connectivity to the central US military network, SIPRNET,[18] and here the UK military has two main routes: one is through the Defence Information Services Network (DISN) and the other is through a variety of bespoke coalition networks called CENTRIXS.[19] The United States recently folded some twenty-seven CENTRIXS into three principal networks: one for global counterterrorism operations (with sixty-nine partners), one for coalition operations in Iraq (with twenty-two partners), and CENTRIXS Four Eyes (US, Britain, Canada, and Australia). DISN offers secure and direct communication to SIPRNET, whereas CENTRIXS offers less security and requires the manual physical transfer of data from colocated CENTRIXS and SPIRNET terminals. Against this, DISN is limited to email and attachments, whereas CENTRIXS can be "ramped up" to provide JOPs.[20]

Both routes of connectivity are cumbersome and hardly provide the network capability required for coalition effects-based operations. The connectivity problem is primarily a matter of politics and procedure, not technology. It has to do with US nervousness about sharing information with allies. Of course, not all allies are treated equally. There are "concentric circles of access," with the British (as the US military's closest ally) enjoying special access rights, followed by Canada and Australia, and then others. Still, most information on SIPRNET is classified NORFORN, meaning "no foreigners," and that includes British military personnel. Given how central SIPRNET has become to US planning and operations, NORFORN classification has created all sorts of problems in coalitions operations in Afghanistan and Iraq, especially when US joint planners put information on SIPRNET without realizing that coalition partners would be unable to access it. It has also led to quite farcical methods to get around such restrictions, such as US officers calling up information on SIPRNET in the presence of British colleagues and then pretending to look the other way.[21] Pressure has come from some US commands, including Joint Forces Command and Strategic Command (STRATCOM) for improved coalition networking.[22] But as of 2009, the British still do not have direct access to SIPRNET.

In terms of developing their own network capability, the British have adopted a less technocentric and plainly less ambitious approach than the United States. Hence, whereas the US military talk of network-centric warfare (NCW), the British speak of network-enabled capability (NEC). In the United States new information technologies are seen as revolutionizing the conduct of operations, in particular, by increasing transparency in battlespace and reducing operational risks. Future US joint forces are expected to be "knowledge empowered" through networking, and hence able to exercise superior speed, agility, and precision in achieving desired effects.[23] This perspective is informed by the impact of IT on the global economy. Just as IT has transformed the way we transact in society, especially in the West, so it is claimed to hold the potential to transform the way we fight.[24] In this sense, the old emphasis on weapon platforms and large standing formations is associated with the industrial age, and contrasted with networks and agile, mission-orientated groupings.[25] Proponents of NCW are quick to note that it is not just about developing new technology, but also about developing the doctrine, organizational structures, training, and practices for network warfare.[26] Nonetheless, this remains a technocentric vision of future warfare—one involving greater automation and a reduced role for human agency. This US vision extends to a hugely ambitious program to create a Global Information Grid (GIG) integrating all US military and Department of Defense (DoD) information systems into one seamless and reliable super-network. The GIG core capabilities are supposed to be established by 2011 at a cost of around $34 billion, with full network development around 2020.[27]

The British simply lack the resources to invest in military networking on this scale. Indeed, the MOD openly acknowledges that even developing a more modest network-enabled capability "in a single bound is prohibitive," given cost and complexity. Hence it has adopted an evolutionary approach involving three phases or "states": an initial NEC state in 2012 improving the interconnections between current equipment, a transitional state in 2017 with improved integration provided by new equipment capability coming into service, and a mature state around 2030 characterized by synchronization of joint military capability.[28] Nonetheless, the British view on networking of forces has been informed by American ideas. Indeed, the US Office of Force Transformation (OFT) began discussions with NATO allies in 2002 about the benefits of networking, and this included NCW briefing visits by the then director of OFT, Admiral Arthur Cebrowski, and his deputy, John Garstka, to the UK Ministry of Defence.[29]

In 2003 a team was established within MOD to develop concepts and doctrine on NEC—in essence, to "translate NCW into NEC." As Deputy Chief of the Defence Staff for Equipment Capability, Lt. General Robert Fulton took charge of the embryonic British NEC effort. Fulton's view was that NEC would impact on virtually all areas of capability development in the MOD. Hence, rather than concentrate NEC in specific equipment programs, Fulton decided to embed NEC across the entire MOD Equipment Plan.[30] This approach continues to define the evolving British NEC program. The MOD's latest assessment is that "nearly 60% of the current 500 or so projects in the equipment plan could be described as significant contributors in some way to NEC."[31] That said, the UK NEC backbone is provided by a few key programs, including the Defence Information Infrastructure (DII), the Bowman secure land tactical communication system, and the Skynet 5 satellite communication system. Bowman has entered service, and all frontline brigades have been "Bowmanised."[32] Bowman promises to speed up operations by providing secure data and voice communications and thereby negating the need for code books. Observations from training confirm that Bowman "increased the tempo of battlegroup operations."[33] However, operational experience in Afghanistan has revealed some shortcomings—namely, limited battery life, limited range in mountainous terrain, and the weight of the unit, which is not ideal for dismounted infantry.[34] UK military satellite communications are currently based on the Skynet 4 system, with two of the three satellites launched in the late 1980s still in service. Skynet 5 entered full service in May 2007, with the first of three satellites in orbit. The new satellites are accessible via the existing MOD ground terminals, but the £3.6 billon Skynet 5 program does also include deployment of a range of new mounted and portable terminals.[35] In short, on the equipment side of NEC, the United Kingdom is making progress, albeit with the usual delays and glitches that plague any large programs of this nature.

Two other factors, beyond resource constraints, are shaping the British approach to NEC. Both are rooted in British military culture. The first is the British military's command culture. There is a natural tension between networking of the force structure and the philosophy of Mission Command, which was adopted by the British military over two decades ago. This philosophy explicitly recognizes the essential human-centric nature of warfare and values the human ability to respond in flexible and even innovative ways to complex and unexpected situations. Thus Mission Command seeks to strike a balance between direction and delegation by encouraging commanders to tell their subordinates

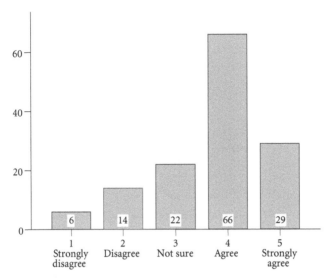

Figure 3.1. The application of information and communication technology will profoundly change the conduct of military operations.

what needs to be achieved and why, and then to let subordinate officers get on with determining how to achieve their commanders' intent. To be sure, information networks offer the possibility to strengthen lateral command practices. But the British military is also acutely aware that by offering senior commanders a clearer picture of what is happening in the joint operating area, networking may tempt them to take out the "long screwdriver" and direct operations on the ground.[36]

A second shaping factor is a natural British military skepticism about what the technology can achieve. It is tempting here to draw a contrast with the technocentrism of US strategic culture.[37] However, it would be a mistake to read any technophobia in British strategic culture. After all, the British were the first to embrace the military possibilities of two key technologies in the interwar period: the airplane and the tank. Britain was the first country to develop an independent air force, and, following the slaughter of World War I, the British Army came to the view that future battles "had to be won at least cost using the maximum of machinery and the minimum of manpower."[38] Hence, in a survey conducted in 2007 of British military officer students at the United Kingdom's Joint Services Command and Staff College,[39] 70 percent of respondents expected that the application of IT "will profoundly change the conduct of military operations" (see Figure 3.1).[40]

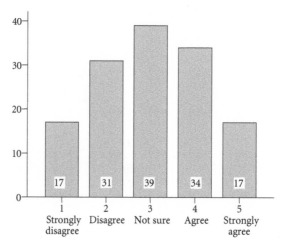

Figure 3.2. Information networks will be the most important asset to the success of military operations in 2020.

However, that is not to say that British officers endorse a network-centric view of things. Hence, respondents were fairly evenly split on the proposition that "information networks will be the most important asset to the success of military operations in 2020" (see Figure 3.2). We conclude from this that while it would be wrong to ascribe a technophobia to the British military, there is certainly a techno-skepticism when it comes to what networking may be expected to achieve.

This skepticism has been reinforced by recent operational experience. The British view is that far from validating NCW, the 2003 Iraq campaign has revealed its limitations. The British commander in Operation *Telic* (the British codename for the 2003 Iraq campaign), Air Vice Marshall Brian Burridge, found that even with access to American ISTAR assets, he lacked the network capability to conduct an effects-based campaign.[41] Similarly, the British Army's confidential lessons-learned report on *Telic* concluded that while NEC provided some positive benefits, coalition victory was secured "largely for old-fashioned reasons": speed, surprise, and superior war-fighting.[42] There is strong anecdotal evidence in the public domain to support this finding. Network capabilities, especially the Blue Force Tracker system, provided a greatly enhanced picture of the location of coalition forces to US commanders at corps headquarters and above, but the network failed to deliver a reliable picture of the disposition of Iraqi forces. This problem was especially acute for commanders in the field, with bandwidth restraints slowing down the joint tactical network and routinely causing it to crash.[43] More recent British operational experience points to the benefits of NEC. Thus, for example, the campaigns in Iraq and Afghanistan have pretty much ended debate within British military circles about the urgent need for unmanned aerial vehicles (UAVs), another key NEC asset.[44] More to the point, Afghanistan illustrates how it is the combination of assets (and not a single new asset) that produces network enablement: hence, Reaper UAVs have been used to generate full-motion video intelligence that has been distributed to British forces in Afghanistan via Skynet 5.[45] This appreciation of operational experience reinforces the cultural predisposition on the part of the British military to reject a full-blown NCW model in favor of a more modest NEC.

EFFECTS-BASED APPROACH TO OPERATIONS (EBAO)

The proclaimed aim of the UK's EBAO was deceptively straightforward: "the way of thinking and specific processes that, together, enable both the integration and effectiveness of the military contribution within a Comprehensive Approach (CA) and the realization of strategic outcomes."[46] This "clarity," however, masked a significant degree of conceptual controversy on both sides of the Atlantic. The inclusion of a section on "delivering strategic effect" in the 2003 DWP gave a publicized green light to the UK doctrinal community to develop what was then known as effects-based operations (EBO) with a view to producing new formal doctrine.[47] The work developed two linked strands. First, how to focus military activity in a manner that kept the strategic purpose of operations at the forefront of the planning and implementation processes. Second, how to coherently situate military lines of operation within a broader diplomatic, political, and economic policy framework.

Timelines are vital in understanding how EBAO thinking in the UK developed. While, of course, there was a general awareness among doctrine aficionados of the development of EBO in the United States, serious concept development toward a new UK doctrine began in earnest only through 2003 and into 2004 in the Joint Doctrine and Concepts Centre (JDCC; now the Development, Concepts and Doctrine Centre—DCDC).[48] Doctrinal work on EBAO developed against the background of a rapidly deteriorating situation in Iraq. Moreover, Iraq suggested that while military action may well be necessary for future British foreign policy objectives, it was increasingly unlikely that such action would be sufficient. The question raised for the UK military, therefore, was how do you bring military lines of operation within a wider, coherent policy framework to achieve national strategic objectives? The perception within DCDC was that a "doctrinal gap" had emerged.[49]

Initial work on an effects-based doctrine in the UK was not helped by the trajectory that concept development appeared to be taking in the intellectual home of effects-based thinking, the United States. The concept of EBO had a much longer history in America. Consequently, while the catalyst for concerted effects-based thinking in the UK was events in 2003 Iraq, the pivotal crisis in the development of EBO thinking in the United States was Operation *Desert Storm* in the 1991 Iraq War—in particular, the perception that airpower had delivered, and could again in the future deliver, spectacularly successful strategic outcomes. The pivotal lessons arising from *Desert Storm* were seen as illustrating the potential for EBO to underpin future military planning with airpower in the vanguard.[50] Significantly, this complemented Defense Secretary Donald Rumsfeld's vision of a military transformation process designed to fully harness the technological advances heralded in NCW to produce a much leaner, more efficient military that could deliver combat success at much lower numerical levels of platforms and, in particular, manpower.[51] The direction of US transformation seemed geared very much toward high-technology, rapidly deployed, short-duration combat missions in which victory could be achieved quickly and forces rapidly withdrawn. This suggested a preference for a highly militarized operational level with the minimum "interference" from civilians in general, and other government departments in particular.

The picture in 2003, therefore, was a less than appealing one from a British perspective. This seeming American preference for viewing the operational level as a depoliticized realm, one that privileged technologies (likely to intensify British-American interoperability issues) and prescribed a mathematical modeling of battlespace, was considerably at odds with the traditional UK approach. Indeed, such a vision was antithetical to a number of strands within British military culture. As noted earlier, there is a streak of techno-skepticism that derives from a strongly held belief that military campaigning is, ultimately, an art as opposed to a science.

Emerging American EBO doctrine also conflicted with the doctrinal approach that has dominated British military thinking since the mid-1990s—namely, the Manoeuvrist Approach. This was first formally articulated in *British Defence Doctrine* in 1996, where it was defined as an approach to operations "in which shattering the enemy's overall cohesion and will to fight, rather than his materiel, is paramount."[52] This approach is realized through the philosophy of Mission Command, which is intended to give British forces agile command and superior operational tempo, enabling them to destroy the enemy's cohesion and will to resist. Success is, thus, delivered by the flexibility inherent in Mission Command and the subtle application of operational art rather than technology, per se, or staff-centered battlespace modeling and direction. These are powerful cultural features of how the British military officer views the nature of operational command. Consequently, a British EBAO, to be accepted, needed to incorporate rather than replace the cultural elements that underpin the Manoeuvrist Approach.[53] It was clear that JDCC would be unable to sell the "matrix-management" approach to British commanders even if it wanted to (which it decidedly did not).

The first formal Joint Doctrine Note on effects-based thinking to come out of JDCC/DCDC appeared in September 2005 entitled JDN 1/05, *The UK Military Effects-Based Approach*.[54] An initial point of interest was that the wording "EBO" had now clearly been jettisoned in favor of "EBA," which was held to reflect more accurately the focus on the appropriate "mind-set" and distanced concept development from the negative connotations applied to EBO through the experience in the US. The document itself was relatively short, and stayed mostly at the conceptual level. Emphasis was placed on the role of the military within an integrated policy framework that potentially includes other government departments, international organizations, nongovernmental organizations, and other nations. Christened the "Comprehensive Approach," this wider policy framework generated a Joint Discussion Note of its own in January 2006.[55] The other key points in 1/05 were relatively uncontroversial precisely because many of its elements were kept at a fairly abstract level. The call for a shift in mind-set away from a central focus on the organization of military activities toward a focus on desired outcomes and end-states was unthreatening because it was broad enough to allow those with command experience to convince themselves that it was not really anything new.

Joint Doctrine Note 7/06 was a much more detailed document, particularly in terms of its focus on effects-based campaigning. It contained specific sections on analysis, planning, execution, and assessment, including a hypothetical campaign vignette complete with basic planning schematics. Of most significance were the implications for traditional British approaches to campaign planning. These are revealed most starkly in the campaign planning schematics that illustrate a hypothetical generic scenario. A "horizontal" relationship is identified in which operational level planners need first to establish the desired *operational end-state*; then delineate the *decisive conditions* that must be achieved to con-tribute to that end-state; followed by outlining the *supporting effects* required to realize those conditions: only then concentrate on the *activities* that may lead to those supporting effects.[56] This is what is meant by privileging outcomes as opposed to activities.

The full implications are revealed, however, only when these generic "boxes" are populated in the campaign vignette.[57] The operational end-state is denoted as being "a lasting peace in which the threat of violence and civil war has been removed and Country 'X' has mature political structures, supported by reli-able infrastructure and governance providing prosperity and security for all its people."[58] This expands the concept of the operational level well beyond what military planners have traditionally considered it to be, and graphically indicates the belief (at least within DCDC) that the operational circumstances in which the UK military are likely to find themselves will place a premium on the ability to coherently integrate military lines of operation in a much wider policy framework.

Concurrently, through 2005 and 2006, events began to shift thinking on EBO in the US in a direction more congenial to those writing UK doctrine. The unraveling of many of the assumptions that shaped the initial US military planning in Iraq encouraged many who harbored doubts about the direction of EBO development to speak out robustly. The extensively modeled, highly predictive elements to EBO began to come under serious intellectual pressure. Marine Corps Lieutenant General James Mattis was particularly vocal in his condemnation of an approach that he dismissed as reducing operational art to a set of equations and algebraic predictions.[59]

There appeared to be further positive developments, from a British perspec-tive, when the senior mentor at the Joint Warfighting Center (JFC) of US Joint Forces Command (USJFCOM) produced a paper distilling "best practices" gleaned from the JFC's deployable training team's visits to US joint headquar-ters around the world. This received an effusive welcome in British military circles.[60] A number of aspects of the paper were guaranteed to receive critical acclaim in the UK. For example, while confirming the utility of the Political, Military, Economic, Social, Infrastructure, and Information (PMESII) systems construct, the paper added a caveat. "Now, some argue that our adversary can be precisely defined and modeled through this system understanding—and we can predict its behavior. We disagree. We believe today's complex environment is far too complex for reliable modeling and prediction of outcomes."[61] Similarly, the paper emphasized what it called "commander-centric operations." The heavily centralized and "staff-centric" aspects that seemed to be implicit in early EBO conceptualizations in the US were comprehensively rejected in the paper, which advocated that commanders "use mission type orders" and "decentralize to the point of being uncomfortable."[62] On effects-based thinking itself, the paper commented that "we've seen effects-based thinking *misinterpreted* by those who *incorrectly* want to over-engineer it and turn it into a bunch of equations, data bases, and check lists" [emphasis in original].[63] All this combined with heavy emphasis on the need to adopt a holistic, multiagency, and multinational approach that moves thinking beyond the traditional notions of the military battlefield has ensured that the paper was assiduously promoted by DCDC.[64]

With USJFCOM and DCDC seemingly on a convergent path with their mutual understanding of EBAO, an opportunity appeared, for the first time, to some transatlantic doctrinal coherence on effects-based thinking. However, the appointment of General Mattis (promoted to four-star upon appointment) to succeed General Lance Smith as JFCOM commander reignited debate in the US about the utility of EBO. Given Mattis's noted opposition to the original way that EBO developed in the US, the obvious question was whether the direction indicated in General Luck's paper would be maintained.

The world did not have long to wait for the answer. In July 2008 a new Luck paper was released that contained a revealingly entitled section: "The Move Away from Effects-Based Operations."[65] This was followed in August 2008 by an internal JFCOM memo circulated by Mattis that comprehensively sounded the death knell for EBO—at least in terms of future JFCOM thinking.[66] The sentiments expressed are clear and unambiguous. "The underlying principles associated with EBO, ONA [Operational Net Assessment] and SoSA [Systems-of-Systems Analysis] are fundamentally flawed and must be removed from our lexicon, training, and operations. . . . Effective immediately USJFCOM will no longer use, sponsor or export the terms and concepts . . ."[67]

Much of the reasoning articulated to explain this move mirrors the arguments Mattis had been making, with increasing gusto, for a number of years. There were, however, two additional elements that are noteworthy. The first suggested that widely differing interpretations of EBO were spreading confusion among allies. Clearly, to take one example, the British articulation of EBA was very different from the US Air Force understanding of EBO. Furthermore, NATO was beginning to interpret EBAO as more akin to the British "Compre-

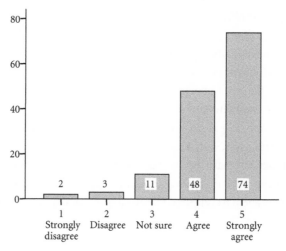

Figure 3.3. Future operations will be characterised by a holistic approach, involving a mix of military and non-military instruments and directed towards achieving strategic ends.

hensive Approach." The second additional element to Mattis's argument drew attention to the Israeli-Hezbollah conflict in the summer of 2006, which, he claimed, graphically illustrated the failure of EBO concepts.

For the British, although they were kept abreast of the way JFCOM thinking was heading, the total rejection of EBO would appear, at first glance, to present a major problem. DCDC, in August 2008, was approaching the end of a major doctrinal rewrite known as the Joint Operations Project, which entailed the redrafting of all the main British campaigning planning and execution doctrine.[68] However, those in DCDC charged with writing the new doctrine are extremely relaxed about the JFCOM approach. Indeed, the jettisoning of the extensively modeled, highly predictive EBO (which is how they interpret the Mattis memo) is viewed as largely positive. The British perspective is that while they will have to "detune" some of the effects-based language and avoid referring explicitly to an effects based approach out of deference to JFCOM sensitivities, the underlying philosophy of EBA as it has evolved in British doctrine can and will be maintained.[69] This is illustrated by the inclusion in the new JDP 5-00 of a planning schematic that reiterates the operational end-state-decisive conditions-supporting effects-activities delineation. Thus the term "campaigning" is being used to describe the overall process with "conditions-based" or "outcome-focused" the preferred descriptive terms for the underpinning philosophy. These are not seen by DCDC, however, as major changes of substance, and have provoked little tension.

Military doctrine is not always as influential as doctrine writers would like. But there is evidence of broad support in the UK military for the underlying philosophy of EBAO. In the survey conducted in support of this chapter, nearly 90 percent agreed with the proposition that "future operations will be characterized by a holistic approach, involving a mix of military and non-military instruments, and directed towards achieving strategic effects" (see Figure 3.3). This is not to say that military officers think force is going out of fashion altogether. Indeed, 80 percent of respondents in our survey agreed with the proposition that "the application of military force will remain as central to military operations in 2020 as it is today" (see Figure 3.4).

It is also clear that EBAO thinking is shaping British plans and operations. It has been incorporated into a core organizational routine—namely, the planning methodology (called the Joint Operational Estimate) that is both taught to, and used by, joint planners at the UK's Permanent Joint Headquarters (PJHQ).[70] Hence, it is entirely credible to claim that EBAO now "lies at the heart of PJHQ."[71] EBAO was also clearly evident in 52 Brigade's campaign when it took charge of Task Force Helmand from October 2007 to April 2008. The campaign plan was focused on generating those effects and decisive conditions necessary to produce an operational end-state, defined in broad terms of secu-

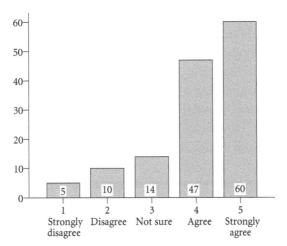

Figure 3.4. The application of military force will remain as central to military operations in 2020 as it is today.

rity, political stability, and economic sustainability. Hence, the center of gravity was conceptualized in terms of the local population rather than insurgent's will and ability to fight.[72] The overarching theme of the campaign plan was "dynamic influence," which meant that military activities were designed to deliver effects tailored for local conditions in different areas of Helmand. To support his plan, the brigade commander concentrated, in particular, on developing the tools to create nonkinetic effects in his campaign. The military have a well established tool-kit to generate and assess kinetic effects (that is, firing weapons and the Battle Damage Assessment system).[73] EBAO is about integrating kinetic *and* nonkinetic effects (such as, "reassure," "influence," and "inform" target groups). Hence, 52 Brigade developed a new methodology (the Tactical Conflict Assessment Framework) to target and measure nonkinetic effects. The brigade also developed two-man nonkinetic effects teams (NKETs), which were deployed down to company level to help deliver and assess nonkinetic effects, as well as larger Development Influence Teams (up to twenty strong), which provided a deployable task force level asset to conduct more in-depth assessments on the ground.[74]

Much more disappointing has been progress in developing the Comprehensive Approach. There is a FCO-led Comprehensive Approach Working Group that is charged with taking forward a cross-governmental approach, but, for the reasons noted earlier, it appears to have made little practical headway. Accordingly, the British military's campaigns in Iraq and Afghanistan have involved ad hoc civil-military cooperation rather than a comprehensive approach. Reflecting on operations in Iraq from July 2005 to January 2006, 7th Armour Brigade noted the excellent working relations with the British consulate in Baghdad. But that was in the context of an overall negative report on poor civil-military linkages in the campaign. Thus, it was also noted that "differences between FCO and MOD were often felt even at the tactical level." The lesson learned from this was that "[t]he disconnect between the various arms of government has undermined faith in the Comprehensive Approach and whether it can be delivered."[75] This view is echoed by 52 Brigade's experience in Afghanistan: the Comprehensive Approach was seen as a Whitehall concept that had no actual impact on the ground in Helmand.[76]

This problem notwithstanding, the "force multiplier" effect of a coherent interagency policy and operations framework is generally attractive in an era of resource constraints, but there are inevitable clashes of departmental and governmental/nongovernmental cultures associated with attempts at wider policy coordination. Implementation of an effects-based approach may generate the occasional frictions, but the philosophy underpinning the Comprehensive Approach and the importance of influence operations is here to stay for the foreseeable future.

EXPEDITIONARY WARFARE

The link between expeditionary capability and strategic effects is clearly identified in the 2003 DWP. Hence, it is noted that "[in] order to deliver a wide range of effects, we need to be able to deploy and configure forces rapidly."[77] In a hearing into the DWP, the House of Commons Defence Committee reported that the white paper had been criticized for lacking detail and having "no real meat." It cited Lord Guthrie, a recent former Chief of Defence Staff, as describing the DWP as "a bland document" that was "full of buzz words and platitudes."[78] This was fair criticism when the DWP is compared with the far more substantial 1998 SDR. It is also understandable in light of the new concepts—especially NEC and EBAO—that appeared in the DWP before they were fully developed, let alone disseminated, in the wider UK defense community. But in terms of expeditionary capability, this emphasis may be traced back to the SDR.

The SDR was much touted as a "foreign policy led" defense review. As noted earlier, the SDR was intended to align future force posture with the foreign policy priorities of the incoming Labour government.[79] Chief among these priorities was the development of an "ethical dimension" to British foreign policy. That, along with the responsibilities of being a permanent member of the UN Security Council, led Labour to seek a more interventionist role for the UK armed forces. Hence, the introduction to the SDR declares that "[we] do not want to stand idly by and watch humanitarian disasters or the aggression of dictators go unchecked." Rather, the aspiration was for Britain to be "a force for good" in the world. Added to this moral imperative for greater British interventionism is a new strategic logic. Simply put in the SDR: "[In] the post Cold War world, we must be prepared to go to crisis, rather than have crisis come to us."[80]

The national security imperative for British military interventionism gained added salience following the 9/11 terrorist attacks on the United States. In response, the government conducted a mini-defense review resulting in the publication of *A New Chapter* to the SDR in July 2002. Citing both the logic of striking at time of one's own choosing, and the logic of deterrence, *New Chap-*

ter advocated power projection against terrorists abroad and state-sponsors of terrorism: "Experience shows that it is better where possible, to engage an enemy at longer range, before they get the opportunity to mount an assault on the UK." Quite reasonably, therefore, *New Chapter* concluded that "if anything, the trend (which we recognised and planned for in the SDR) towards expeditionary operations—such as those in recent years in the Balkans, in Sierra Leone, in East Timor and in and around Afghanistan—will become even more pronounced."[81]

Obviously, the new emphasis on expeditionary warfare has had significant implications for navy and army force development. So far, the RAF has not really been affected. There have been some logistical adjustments in order to provide greater frontline support to combat aircraft.[82] However, major investment in airlift and ground support capability, which are crucial for expeditionary operations, is unlikely. For starters, the RAF does not consider that its core role is to support the land battle. In any case, there has already been massive investment in the RAF's prestige legacy program, the Eurofighter (now designated Typhoon). Around £19 billion is being spent acquiring 232 Typhoons, which have been retrofitted to enhance their capability for air-to-ground operations.[83] Arguably, some of this money might have been better spent on tactical airlift, which is in short supply in the current British campaign in Afghanistan.[84] Moreover, the existing airlift fleet is in urgent need of modernization.[85] But the money has been spent.

Equally, the Royal Navy has got what it has wanted—namely, two new large-deck aircraft carriers. But in this case, the post–Cold War emphasis on expeditionary warfare has underpinned the case for the new carriers. Indeed, this requirement was stated up front in the SDR.[86] To give some idea of the increased capability these will bring, the existing 20,000-ton Invincible class carries twenty-two aircraft, whereas the Future Carrier will be 65,000 tons and carry forty aircraft. That said, the new carriers are conventionally powered, unlike the nuclear-powered US large-deck carriers, and this means that fuel is taking up much room that otherwise would have been available for additional aircraft and equipment. Also they are going to operate a ramp launch system, consistent with the existing Invincible class, instead of the catapult system used by the US Navy. This creates limitations both in terms of interoperability with the US Navy, and also in terms of requiring the Royal Navy to select a short take off and vertical landing (STOVL) replacement for its aging fleet of Harrier strike aircraft. The Royal Navy has selected the STOVL variant of the F-35, with a planned purchase of 138 F-35s and an in-service date of 2017. However, this program has experienced both design and schedule problems, and one may anticipate further complications that may delay entry into service. At the same time, the MOD announced in December 2008 that the in-service date for the two new carriers was being put back by two years—to 2016 and 2018, respectively—because of pressure on the defense budget. Combined, these two programs will take up the lion's share of the Royal Navy's procurement budget and so signal a major investment in expeditionary capability: the Future Carrier is expected to cost £4 billion for two vessels, while the F-35 is anticipated to cost in the region of £7–10 billion.[87] Here, too, it might be argued that money would have been better spent on acquiring a larger number of smaller carriers, or indeed on more ambitious assault ships, along the lines of HMS *Ocean* (which displaces 21,000 tons and carries an aviation wing of eighteen helicopters).[88]

Of all the services, the army has been most affected by the post–Cold War adjustments. The primary mission of the army shifted from forward defense of NATO by armor-heavy divisions based in Germany, to expeditionary operations by mechanized infantry battlegroups in Iraq and Afghanistan. Already this change was anticipated in the SDR with the creation of the Joint Rapid Reaction Force, to which was assigned the army's elite deployable brigades (the then 24 Airmobile, and ready brigades from 1 and 3 Armoured Divisions), as well as 3 Commando Brigade from the Royal Marines.[89] The main effort since SDR has focused on "rebalancing" the army for the post–Cold War world. DWP establishes the requirement for more capable deployable brigades and lower level battlegroups, especially in terms of logistics and other "force enablers." It also sets the new 2-3-1 Future Army Structure—that is, two heavy armored brigades, three medium-weight brigades, and a light brigade (a second light brigade has since been added).[90]

We should note that the move to medium weight was not forced on the army by civilian policymakers. To be sure, the ability to field a war-fighting division remains the gold standard for the British Army, not in the least to maintain its credibility especially with the Americans. But the army itself recognizes in its *Army of Tomorrow* study that "we are likely to find ourselves more frequently deploying rapidly for operations at brigade and battlegroup level, followed by enduring operations at the same scales of effort."[91] The experience in Iraq since 2003 and Afghanistan since 2006 (when the British Army deployed in force in the south) conforms to this prediction. Moreover, the types of operation provided in the FLOC are all expeditionary in nature.[92]

The legacy force, with its light-heavy force mix, was not optimized for expeditionary operations. Hence, the army's transformation plans have centered on developing a new medium-weight capability. Army chiefs have understandably focused on the new ground maneuver platform that would deliver this capability: the Future Rapid Effects System (FRES). The FRES program is innovative in form. Whereas previously the army has had single vehicle programs, such as the Warrior infantry fighting vehicle (IFV) and the Challenger 2 main battle tank (MBT), FRES is sixteen vehicle variants based on a few generic platforms. Clustered in two families—utility and specialist—these variants are expected to deliver the full range of medium-weight battlefield functions including protected mobility, medium armor, reconnaissance, command and control, combat support, and combat engineering. FRES has undergone considerable changes since the original concept was outlined in 2001. The number of vehicles to be acquired has increased from approximately 1,500 to well over 3,500, and the vehicle weight has increased from 17 tonnes to 25–32 tonnes. The weight increase is significant, for it has required abandoning the original requirement for FRES to be transportable by a C-130, which would have given FRES added in-theater mobility. However, learning lessons from current operations in Iraq and Afghanistan, the army has determined that additional armor is more important than greater transportability.[93] Moreover, in terms of strategic airlift, this requirement will be met by the acquisition of twenty-five new A400Ms that have a revised in-service date of 2012–13, as well as the enlarged fleet of six C-17s.[94]

FRES was originally supposed to be fielded by 2009. There is considerable confusion now over when, and in what number, FRES will enter service. The Army's Fleet Review in December 2005 identified two possible dates: 2017 for a fully developed program, and 2012 for an early but suboptimal program. The army initially selected 2017, and that date was reported to the Commons Defence Committee in early 2007.[95] However, shortly thereafter, the army moved to the earlier date following intervention by the then defense procurement minister, with effort being focused on developing the utility variant.[96] In November 2008, it was reported that because of an inability to conclude contracts with General Dynamics to develop the FRES utility platform, the whole program was being restructured.[97] In the past two years, Britain has urgently acquired more than 280 new armored vehicles to counter the IED (Improvised Explosive Device) and mine threat to combat forces operating in Afghanistan and Iraq.

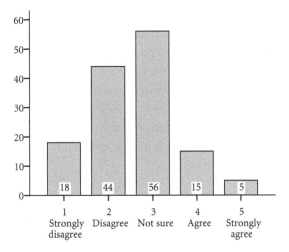

Figure 3.5. NATO's Allied Command Transformation is a crucial driver of military transformation in your own country.

In October 2008, the government also announced a Protected Mobility Package to acquire 700 additional vehicles by mid-2011 at a cost of £700 million, to provide similar armored and counter-IED mobility to logistics and other enabling units operating in the field with combat troops. All these are off-the-shelf acquisitions—that is to say, existing US-produced vehicles are being bought and enhanced to meet UK requirements.[98] The view in the UK Treasury is that with hundreds of new armored vehicles in the field force, this reduces the immediate imperative to develop the FRES utility platform. Hence, it is likely that a restructured FRES program will concentrate on developing the specialist range of vehicles.

Clearly, the changed strategic environment is the major driver in the development of expeditionary capabilities. Indicative of how states and institutions have responded is the reinvention of NATO as a global security organization for the post–Cold War era. Especially significant here was the 2002 NATO Summit in Prague, where it was agreed to establish Allied Command Transformation (ACT). As noted in Chapter 1, ACT was specifically designed to be a driver of military transformation in the alliance. However, there is little evidence that ACT is playing much of a role in shaping British military transformation. Certainly our survey results indicate that few British officers think that ACT is an important agent of transformation in their own country (see Figure 3.5). A more significant driver for the British is the need to keep step with the Americans. The bottom line is that the British expect to be doing major expeditionary operations with the United States, if not under US leadership, and so it is "to [the US] tempo of operations that the UK must aspire."[99] The conclusions for the development of British expeditionary capability are that "UK forces therefore need to remain closely aligned with US thinking and force development, notably through the pursuit of interoperability and compatibility, as well as close engagement with the US 'transformation' process, advanced technologies, and innovative expeditionary design."[100]

CONCLUSION

The British military is embarked on a comprehensive process of transformation toward a network-enabled, effects-orientated, and expeditionary force posture. This process involves both the development of new doctrine and concepts on NEC and EBAO, and a full portfolio of equipment programs for network enablement, as well as the new large-deck carriers and new medium-weigh armored vehicles.

The drivers of British military transformation are external. Chief among them has been the changed strategic landscape following the end of the Cold War. The collapse of the Soviet threat removed the primary mission of the British armed forces, and necessitated discovery of, and readjustment to, a new set of missions. This involved return to what is, in historical terms, the traditional mission-set of the British military—namely, expeditionary warfare. The other main external driver has been US military transformation—both as an imperative, in the constant concern of the British military to be able to fight with the Americans, and as a source of ideas, especially about the networking of forces and effects-based thinking.

At the same time, the British have not simply aped their US ally. Rather, they have taken American ideas and adapted them to suit British circumstances and sensibilities. Two domestic factors, in particular, have shaped this process: resource constraints and military culture. Britain cannot hope to match, even in relative terms, the level of US investment in military transformation. Thus, even if they wanted to, the British military could not aim for a US-style, wholly networked military force. Nor can the British military afford to develop weapon systems incorporating the next generation of technologies. Thus FRES is far less technologically ambitious than the next generation US medium-weight land platform, the Future Combat Systems (FCS). Thus whereas FCS requires the development of a bespoke network that is the most complex IT program in US federal government history, FRES is to be "networked" by integrating a version of Bowman.[101] Reinforcing these resource considerations is British military culture, both in specific aspects (such as the command culture that sits uneasily with complete network-centrism) and in terms of a general skepticism when it comes to the promises made on behalf of new technologies. Hence, Britain is planning to develop a network-enabled rather than network-centric force structure, and acquiring a medium-weight land force that combines off-the-shelf vehicles and new platforms that incorporate fairly mature technologies.

The British military is entering a period of extreme budgetary pressure. The cost of operations and urgent equipment acquisitions for Afghanistan and Iraq has been met from a special Treasury reserve fund. But in December 2008 it put a cap on this fund of just over £600 million for 2009–10 for urgent equipment acquisitions. Any expenditure over this cap has to be repaid by the Ministry of Defence within two years. This move is indicative of a general pressure on the UK government finances. The defense budget is set to rise by 1.5 percent in real terms from £34 billion in 2008/9, to £35.3 billion in 2009/10 and £36.9 billion in 2010/11. However, there was a £2 billion deficit in the defense budget in 2008/9. Moreover, defense programs (especially equipment) typically experience annual rates of inflation approaching 10 percent.[102] Moreover, the pressure on the defense budget is likely to grow ever more intense in the coming years as a consequence of the increased public borrowing to stabilize the UK banking institutions and economy in the wake of the 2008 global financial crisis. Expenditure will have to be cut by an average of 2.3 percent across all departments in order to service the increased public debt.[103] The government will avoid major cuts in spending on public services before the next general election (likely to be in May 2010), but from 2011 on massive cuts may be expected to fall on defense as on other departments.

These budget cutbacks will undoubtedly affect capability development, but they will not significantly alter the direction of British military transformation. NEC is embedded in so many programs in the equipment plan that the increasing networking of the British military is ensured, even if at a reduced pace. The core concepts of EBAO are well embedded in British doctrine, planning routines, and operations. Moreover, defense budget constraints underpin the logic of closer collaboration with civilian partners in an interagency approach to operations. Of all the areas of military transformation, it is the development of the next generation of expeditionary capabilities, especially for the land forces, that will be most adversely affected by budget cuts. In the interim, the British Army have rapidly modernized their protected mobility fleet using the current generation of armored vehicles. Thus even if FRES is postponed for many years, Britain will still have the best equipped and most experienced expeditionary army of any European state.

4 From Bottom-Up to Top-Down Transformation: Military Change in France

Sten Rynning

Charles de Gaulle's anticipation of military change—Germany's organization of mechanized and offensive operations—motivated his repeated calls for French military reform along German lines.[1] The swift French defeat in May 1940, de Gaulle later noted,[2] resulted from a lack of "willpower," which was precisely what de Gaulle later, as president of the Fifth Republic, put into the making of a new military strategy centered on nuclear weapons.

We detect in this history some issues of relevance to this chapter. First of all, current French military reforms aim to move beyond the nuclear deterrence strategy conceived of and embedded by de Gaulle. Military reform in France— the projection of conventional forces outside the "sanctuarized" national space—is therefore also a question of reforming the Gaullist legacy, which is no easy task. Secondly, willpower is at best a political metaphor for the complexity of handling military change and a concept best suited to underscoring the hero in history. De Gaulle, like Napoleon, managed military change—respectively, the making of nuclear strategy and the organization of a national army—but only with the help of already emerging capabilities and the executive authority granted by war. Whether we are dealing with Léon Blum, Edouard Daladier, and Paul Reynaud in the 1930s or François Mitterrand, Jacques Chirac, and Nikolas Sarkozy, the post–Cold War presidents, we detect few heroes and instead complex negotiations among many actors, military as well as civilian.

Yet France is in the midst of a sustained and significant military reform process. The extent of change is notable: the force structure of all services, but notably the army, is new, given the decision to abandon conscription and professionalize the forces; training has changed to suit a new environment marked by new technology and combined and joint operations; new doctrines emerge to match these operations and to attach French forces to French foreign policy goals; and the acquisition of force systems reflects the desire to gain the ability to project state-of-the-art forces.

The French reforms in some ways match the designs for "military transformation" emanating from the United States. Technology is a major preoccupation of the reformers; technology is being put at the service of expeditionary warfare; and there is no end point to the process of transformation. To be sure, military planning involves stages and phases, but these are here to drive the process forward without end.

This type of transformation contains several challenges. There is the question of money, which is perpetually short. There is the question of political ambitions that usually exceed organizational capacity. Most fundamentally, perhaps, there is a question of gauging the changing nature of warfare and adapting to it. New asymmetrical and low-intensity conflicts challenge technologically advanced forces. Some would even argue that the latter forces are irrelevant.[3] This may not be the case, but it is certain that the drivers of military reform—money, people, and ideas—must be put at the service of preparing the military for the complexities of the operations in which it will become engaged. French forces, like all Western forces, must combine kinetics and nonkinetics and gain insights into the thinking of their adversaries. They must also gain a sense of the type of "strategic effects" they can hope to achieve.

France has generally responded to this set of challenges by pursuing transformation while emphasizing the interrelatedness of a short and intense phase of violent conflict to longer and complex phases of nation-building and stabilization. French forces are not out to "conquer" but to serve international law and peace.[4] The following analysis of French military change begins with a brief overview of the post–Cold War years to set the background for current changes. The extent of change is gauged along three dimensions of military affairs—network-centric warfare; expeditionary warfare; and effect-based operations. A final section sums up the trajectory of change and pinpoints the principal challenges facing French forces.

MILITARY CHANGE IN FRANCE

Transformation was officially endorsed by the Chief of Staff in April 2005 with the plan "Document de politique générale sur la transformation," which was finalized only in 2006. It proceeds from a political purpose (to gain influence in general and prepare for the role as a European lead nation in particular) to establish a military purpose (to be expeditionary and interoperable and capable in a combined and joint environment). This 2005–6 plan is not the onset of reform but one of its outcomes, one should note. The Chief of Staff reacted to developments taking place in allied forces, notably American and British, and reform efforts at inferior levels of the French forces. Today, the Chief of Staff is one of the drivers of the process, but this happened in a stop-and-go process that owes much to the French tradition of nuclear deterrence and civilian control that began once the Algerian War came to an end (in 1962).

For nearly three decades French military thinking—*la pensée militaire*—lay dormant: prominent strategists such as André Beaufre, Pierre Gallois, and Lucien Poirier operated outside the military hierarchy and wrote on nuclear strategy, a domain that enhanced civilian control of the military. New thinking was forced onto the agenda following the French experience in Operation *Desert Storm* in 1991. The French force contribution—Operation *Daguet*—of 18,000 troops proved hard to organize and deploy, and it was poorly prepared for the kind of offensive operations (Air Land Battle) for which the integrated NATO forces had trained. This "cruel" experience, a "cultural shock," provoked the French forces to plan for force projection, joint warfare, improved command and control as well as logistics, and to emphasize investments in command, communications, and reconnaissance satellites.[5] However, thorough reform was impossible given François Mitterrand's attachment to nuclear deterrence and national conscription, although conceptual change did take place with the defense White Paper elaborated under the right-wing Balladur government (1993–95).[6]

Real change awaited the election of Jacques Chirac in 1995. By 1996, a new military program law was adopted that emphasized conventional (as opposed to nuclear) force, abandoned conscription, and aimed to create a significant French force-projection capability. The horizon for full operating capability was 2015, hence the reference to a "model 2015," and three program laws were foreseen to realize it (covering 1997–2002; 2003–8; 2009–15). The French headline goal was to be able to deploy up to 50,000 troops in one theater or to deploy 30,000 troops in one theater and 5,000 in another. The reforms had notable effects. Thus, by 2004 four defense analysts concluded: "In both deployed and planned systems, France possesses arguably the most advanced operational battlespace digitization program in Europe."[7]

Problems remained, however. The new unfolding armed forces were professional but insufficiently joint. Change did not qualify as transformation. Political cohabitation (following the election of a Socialist government in 1997) meant that military affairs were lost in saliency, which translated directly into a loss of defense money: the durability of the Jospin government (it sat for five years) and its priorities meant that the program law 1997–2002 "lost" one full year in budgetary terms. Military transformation therefore began bottom-up in France: the political level was frozen into a standoff, and the military services observed and experienced inadequate preparation in allied affairs.

We noted earlier that military change in France tends to happen in a stop-and-go process. President Sarkozy, elected in May 2007, intends to break this pattern and organize the kind of evolving innovation that characterizes transformation. He announced this ambition with his June 2008 White Paper on defense and national security.[8] This White Paper is intended as a "living" document subject to regular updates; it is intended to avoid the kind of dogmatic monuments that previous White Papers have become—notably, the Gaullist White Paper of 1972. The 2008 White Paper also introduced two new issues that we should note here:

- It introduced a hierarchy of defense missions where equality previously had reigned: knowledge and anticipation as well as prevention are now the top two priorities; the bottom priority, no. 5, is intervention and expeditionary warfare. The intermediary priorities are deterrence and protection.
- While this could indicate a downgrading of the expeditionary goal of transformation, that is not necessarily so. The goal is to increase the budget for force modernization by almost 50 percent compared with the 1990s. The improved ratio of capital per soldier comes at an expense for the services: the land forces had to cut 40,000 personnel, and also had to reduce the maximum deployable force from 50,000 to 30,000 troops. However, with more capital, the land forces can pursue transformation and thus the idea of delivering a big and modern bang for the buck that puts France and its land forces on par with British forces.

The next question is whether the land forces are ready and motivated for transformation. They are, but in their own way, which is the conclusion of the following overview.

NETWORK-CENTRIC WARFARE

Network-centric warfare (NCW), which emerged conceptually in 1998–99 in the United States, takes us to the heart of the technological dimension of military change, given the emphasis on issues such as "infostructure," "battle space awareness," "virtual collaboration," and "self-synchronization." The French paper trail, notably that of the army, begins in the wake of the American debate. The transformation of the army became a subject of organizational reflection in late 2000 and early 2001 when the army staff produced the document "Future Engagement of Land Forces."[9] Three years later, in January 2004, the army Chief of Staff, General Bernard Thorette, presented the contours of the project as "Future Land Forces 2025" (FTF 2025).[10] The army's planning office—Bureau de conception des systèmes de forces (BCSF)—then wrote two reports addressing, first, the coherence of capabilities of the land forces and, second, those capabilities in light of the 2025 horizon.[11] This work finally led to the final transformation blueprint, the June 2005 report "Capability Transformation of the Land Forces."[12] Thus, army thinking began well before the presentation of the Chief of Staff's overarching framework and concluded shortly thereafter.

General Thorette's intervention in early 2004 was not terribly specific but did strike the themes inherent in the overall transformation debate. The FTF 2025 is not a fixed structure but a continuous adaptation in the search for superiority in certain military domains (battle space information, air-land maneuver, and long-distance precision strikes). Technology is important but evolving so rapidly that the army must reflect further before defining its new force structure.[13]

This wait-and-see attitude also marks the 2005 transformation blueprint, although it offers some detail on the subject, notably in one of the three so-called constitutive concepts dealing with NCW.[14] Properly speaking, it deals with "network-enabling," that being indicative of a less ambitious strategy than the American NCW. Previous French planning in this domain adopted the American term "information superiority," but the 2005 document discards that term in favor of "network-enabling" and its related greater emphasis on the operational context. "Technology is not everything," as the constitutive concept concludes.[15]

The constitutive concept reads like a road map for the people who will continue the digitization and networking of the land forces. It asks them to ponder three NCW dimensions: desired operational effects; impact within the process of war (that is, on the interaction of units); and impact on the various capacities (armaments systems, people and organization, and doctrines). It provides no fixed answers. For instance, it instructs people to avoid information overload by simplifying and synthesizing information. New doctrines must take into account the greater opportunities for troop dispersal, greater distance between sensor and shooter, self-synchronization, and compressed decision cycles. Organizationally, people should balance the need for hierarchy and the new imperative of networked organizations. And so on.

The most notable contribution lies in its emphasis on network-enabling as opposed to network centrism and the role of people in a rapidly changing technological environment. While there is no doubt that the goal is to develop the type of interoperable joint capability that is also on the agenda in the United States and elsewhere, the world of network-enabling is described as a type of "plasma." Plasma may be an appropriate metaphor, but it is hardly firm planning guidance.

The vagueness of the 2005 concept should not detract from the fact that France has invested significantly in "network-enabling" capabilities. French efforts find their roots in NATO, where so-called Multinational Digitized Interoperability Exercises (MDIE) tested the ability of national forces to operate in digitized combined and joint environments. These exercises, in the 1990s,[16] sparked new thinking in France, where notably the army staff in 1998–99 elaborated a concept for digitization—*Numérisation de l'espace de bataille* (NEB). NEB deals with information technology and systems, and it intends to involve all force levels in one network (previous army systems operated only at strategic levels).

By the time the army staff started reflecting on digitization, between 1998 and 1999, force architects in the armaments procurement agency (DGA) had already started reflecting critically on the implications of the 1996 reforms. The starting point was not intellectual or doctrinal: it was the realization that systems' costs had to be cut by 30 percent.[17] This material constraint emerged in 1997. In reaction the force architects looked at the 1996 reform program and realized the extent to which its acquisitions were stovepiped along service lines: far too few capabilities were joint or planned as such. Financial constraints thus joined the MDIE experience to create a growing incentive for change, for transformation.

Christophe Jurczak, attached to the DGA's air-land division, recalls that in 1998 they put together a working-group of fifteen officers (from all services) and fifteen DGA engineers to imagine what combat missions in 2025 would be like.[18] Jointness was paramount, and the group accordingly developed eight distinct and joint force systems that combined would enable and define France's combat ability in 2025. The working group in fact picked up where policymakers had left a review exercise in 1997: as part of reviewing model 2015, the armed forces had produced an overview of thirty key capabilities for the next thirty years. Policymakers failed to prioritize—they made cuts across the board—but the overview served to create a prospective planning document (PP30) that the DGA-led working group refocused around the theme of jointness. The eight force systems of 1998 thus came about as the thirty key capabilities were bundled into packages.[19] Officials today are careful to underscore how the DGA, building on the model 2015 review, actively promoted a capability approach explicitly designed to advance jointness (and achieve rationalizations).

The capability approach did not easily prevail in French planning. The army staff, reflecting on the MDIE experience, developed an initial NEB concept that was narrowly designed for the French Army only. It was *franco-français*, as one officer recalled.[20] They developed a communications system, *Système informatique de communication de l'armée de terre* (SICAT), that quite reasonably was suited to the army's operational needs as opposed to technological or industrial imperatives: the problem was that SICAT did not communicate outside the army. It was neither joint nor combined, and by 2000 everyone realized this failure of design. Rather than creating interfaces between systems (bound to multiply endlessly), it was better to create a new and truly interoperable system. One might say that two years of insulated NEB reflection prepared the army for a different joint capability approach, which brings us to the architecture unfolding today.

The current ambition in terms of communication, command, and control is to create a "core" or a "pivot" that will provide common services to all armed units, be they French or allied. In France the pivot will not be SICAT, which is destined to disappear, but the navy's system, SIC21, which will be joint. The allied dimension comes in via the NATO five-powers interoperability group consisting of the United States, Great Britain, France, Germany, and Italy. The project of this group, "C4ISR On-the-Move," is tied in with transformation in the US land forces but is also intended to provide interoperability the moment a coalition takes form. The armies of the five nations are setting up (at the time of writing) an army interoperability center in New York City, and France and the United States appear ready to sign a bilateral interoperability agreement akin to the one entered into by the United States and Great Britain, although bilateral agreements may give way to one overarching multilateral agreement. However, the point is that for France the reference now is NATO, and the context now by definition is combined and joint.

This brings us back to NEB and the army ambition to be fully digitized and networked by 2015. Two experimental forces spearhead this evolution: the armored 2nd Brigade stationed at Orléans and the light-armored 6th Brigade stationed at Nîmes. By 2003–4 these two forces were singled out as spearheads, and the goal is to have their units, including support units, fully digitized and deployable as such by 2009. Once there, the remainder of the army's combat forces can follow suit to attain the 2015 goal.

A first NEB exercise, Nemausus, was held in November 2004 in Nîmes, where new NEB technologies were blended into a peace enforcement scenario.[21] The first real test took place in March–April 2006, however, when the exercise Anvil 2006 in central France involved the 6th Brigade and for the first time a complete digitized communications system.[22] Some months later the first theater NEB test took place when several regiments from the 6th Brigade deployed to the Ivory Coast for a four-month period. The 2nd Brigade gained operational experience when thirteen Leclerc tanks and around 900 soldiers in September 2006 were deployed as part of a 1,600-strong French force to southern Lebanon as part of a UN peacekeeping force. While the objective was not merely to test the NEB, the Lebanon engagement provided a real opportunity to do so as part of a larger operation. The NEB has since been tested in 2007 in Kosovo and Afghanistan.

There is in fact a third spearhead, although references are normally made to the above-mentioned two: the French-German Brigade (FGB). According to the FGB "Common Vision," issued in December 2004, the two army staffs intend to make the FGB a core element both of the EU's battle groups and NATO's Reaction Force (NRF), and it is planned to become a "workbench" to carry out NCW tests in doctrine and interoperability.[23] This makes the FGB a de facto test bed for the development of multinational cooperation in NCW interoperability, which, again, must proceed from the identification of a "pivot." France is more advanced in this respect and more aligned with the unfolding NATO architecture, which is why it was the French and not the German system that took over when the FGB participated in NRF7 in the second half of 2006.[24]

With NEB, spearheads, and an incipient communications structure, the army was ready to present its FTF 2025 project (January 2004). As will be recalled, this was followed in April 2005 by the army staff transformation project, containing, among other things, the NCW constitutive concept. These documents reflect lessons learned and will create a new fighting force that comes under two labels:

- Scorpion (*Synergie du contact renforcée par la polyvalence et l'infovalorisation*);
- Future contact system (*Système de contact de futur* [SCF]).

Both are designs for the new army combat force (that is, forces that in one way or the other establish "contact" with the opponent). Scorpion was presented in late 2004 following collaboration between the army staff and DGA, and SCF is the future force still to be defined. Scorpion will be examined below; the point here is that with Scorpion and FTF 2025 the French Army has a transformation blueprint that builds on network-enabling and that will be organized around a combined and joint communications infrastructure developed in close cooperation with leading NATO allies.

EXPEDITIONARY WARFARE

French conventional forces must be of high quality; they must be readily deployable to the areas of concern for France; and they must be ready for combat on their own or with allied forces. That statement would seem obvious today. But it was wildly controversial in September 1975 when Chief of Staff General Méry made it,[25] and when the defenders of de Gaulle's legacy sensed a conspiracy, instigated by the new non-Gaullist president, Valéry Giscard d'Estaing, to turn back the clock on nuclear deterrence. Jacques Chirac was influential in forcing Méry's vision into retreat in the late 1970s, and it is therefore ironic that Chirac, some twenty years later and now as president, embraced the Méry agenda of creating intervention-capable forces. "The world is changing," argued Chirac. "France must change along with the world."[26]

In the previous section we encountered some of the problems associated with the making of this type of force: insufficient jointness; budget overruns; conceptualization, training, and development in complex institutional and political environment. This section will assess the expeditionary nature of FTF 2025 and Scorpion.

The June 2005 document "Capability Transformation of the Land Forces" led to a second "constitutive concept" dealing with expeditionary forces.[27] The new priorities are first-entry, sustainability, and adaptability. Adaptability is really the key-word (that is, *polyvalence*), and the document goes to some length to define what it is and what it is not. Adaptability resides in land forces whose capabilities and training make them ready for multiple roles (that is, *multi-rôles*), flexibly combining fire and mobility in distinct force packages and thus distinct force options for decision-makers. Adaptability is not a question of particular capability platforms, nor is it a renunciation of certain engagements (for instance, high-intensity combat). It is rather the French version of full-spectrum engagement forces.

The French document draws on the general military debate regarding the simultaneous occurrence of high-intensity combat, peacekeeping operations, and humanitarian assistance (a three-block war) and the importance of decisions made by small unit leaders (the strategic corporal).[28] This focus on the pivotal role played by soldiers, as opposed to weapon or communication systems, dovetails not only with the lessons being drawn in France from Afghanistan and Iraq but also with the "operational culture" of the French land forces, according to the Chief of Staff's 2005–6 transformation document. The autonomous and critically reflecting soldier is a necessity but also, clearly, a challenge to the traditions of military hierarchy. The problem is not hierarchy as such but the soldier's identity and esprit de corps that follows from group organization, the army 2005 concept notes. The making of expeditionary forces could potentially lead to a "cultural rupture." Transformation must therefore involve the construction of a new military identity, a new esprit de corps.

The French forces must in this respect build on the reform process begun in the 1990s, which led to the emphasis on expeditionary forces and professionalization in 1996 and which historically builds on the experiences and culture of the force units that always were expeditionary (that is, the foreign legion, the marines, the paratroopers). Historically, these units were limited in numbers and light: the current challenge is to enlarge the numbers significantly while making them heavier. The French emphasis on multiple roles (*multirôles*) reflects a desire to place the center of gravity on forces in the middle of the spectrum, between light and heavy. Thus, for several years following the incipient transformation agenda in 1998–99, up until 2005, the French concept was not *forces multirôles* but *forces médianes*.

There was never any concern that these median forces would be able to include the traditional light expeditionary forces, but would they be able to include regular heavy forces? They had to, was the answer revealed by the decision to begin NCW experimentation with the 2nd Armored Brigade headquartered in Orléans and organized around eighty Leclerc tanks. "There isn't any ideal mix of light, median, and heavy forces," declared one defense minister official, but the idea is to keep options open while focusing on the middle of the spectrum, the multirole forces.[29]

Uncertainty is thus built into the expeditionary forces' adaptability, and the Scorpion program of 2004 is intended to alleviate uncertainty. Scorpion is an integrated program outlining the kind of capacities the land forces need in the phase of "contact" with the adversary, and its great virtue is its visibility and integrative nature. Scorpion presents most of the land forces (that is, only those units in "contact" with the adversary) as a whole, as one big system on par with the navy's de Gaulle aircraft carrier or the air force's multipurpose jet, the Rafale. Without Scorpion, the army would have been vulnerable to budgetary cutbacks because it is easier, as the defense minister official continued, to cut a few radios from the army than parts of an aircraft carrier. However, if these radios quite apparently form part of a ground force about to enter into direct contact with the adversary, it is distinctly more difficult to make budgetary cuts there.

This rationale was emphasized also by most officials interviewed for this article. The army had too many small programs and lacked coherence. The first step toward new coherence took place not with Scorpion but around 2002 with the idea of creating an "air-land bubble" (*bulle aéroterrestre* [BAO]). BAO is the first federating concept for deployed land forces. BAO is similar to a system-of-systems approach insofar as it is an intranet that links air-land units in a theater. It is not only an intranet, though: it is designed around a core capability, the light armored vehicle, which then connects to various capabilities, some of them still in the making (that is, robots, drones, and so forth).[30]

The centrality of the light armored vehicle fits with the multirole/median emphasis discussed above and corresponds to the inclusion of the 6th Brigade in the NCW development (it consists of around 6,000 troops and 2,100 vehicles in six regiments). It also corresponds to the trajectory of American land forces in which brigade combat teams (BCT) are organized around ground maneuver brigades—of which the most reputed is the Stryker Brigade with its new armored vehicle. In the United States, the BCT will eventually become the future combat system (FCS)—the transformed land force. In France, there is as yet no clear link between the Scorpion and the future system (the aforementioned SCF), but there is a similar effort to make ground maneuver brigades centerpieces and to equip them with a new generation of vehicles. These are labeled EBRC (*engin blindé à roués de contact*) and will notably be connected to all units in the BOA.

The light armored vehicles will be accompanied on the one hand by heavy armor in the shape of the Leclerc tanks and on the other hand transformed infantry units or regiments known as Félin.[31] Félin will first aim to connect the soldier to the entire network of the BOA; next it will seek his integration with robots to develop a new type of individual combat. This latter project, labeled HOBOT (*HOmme—roBOT*), is still vague—the horizon is 2015 to 2025—and will be defined along with the FCS.

BOA and Scorpion will in the long run lead to FCS, and it is too early to tell what FCS might look like. We do not yet have an underlying concept for the FCS, as one officer who is in charge of this program told me.[32] To drive the process forward, though, the DGA has become host to a new battle-lab (LTO),[33] situated in the suburbs of Paris. It was inaugurated in October 2006 and is intended to help find solutions to operational needs at system-of-systems levels. This is to say that the LTO is there to help the services, the DGA itself, and the defense industry reflect on jointness. The LTO can work with any joint system—missile defense, for instance—which is to say also BOA and Scorpion. The LTO is too young to be widely known in the system of the armed forces, as one LTO official argued, but in time it should promote thinking across service cultures.[34] LTO combined with BOA and Scorpion developments could help the army shape the revised esprit de corps called for in 2005. However, the LTO is anchored outside the army, and the army maintains its own centers of development and research for nonjoint purposes; there is likely organizational friction to be overcome before the LTO-army relationship operates smoothly.

The French defense industry is important to the LTO. In December 2005 three companies—Thales, Giat, and SAGEM—were awarded a contract by DGA regarding not only the LTO but also BOA and its system-of-systems. This phase of initial implementation is proceeding with the usual problems of coordination and clarification. At present there are two competing systems for the future SCF, and quite obviously, at some point, someone must choose a winner because there can be only one system-of-systems. The choice is partially for the DGA and the army to make, but the industrial interests involved will politicize it. Moreover, Scorpion is to achieve initial operating capability by 2008, while the BOA system-of-systems is set to achieve the same status four years later, by 2012. Clearly these will have to be better coordinated, and if industrial competition delays BOA, or at least prevents its advancement, it looks like Scorpion will have to be delayed.

The BOA-Scorpion-SCF process has drivers outside the world of force planning. These reside notably in the overall organization of the French forces. The 1996 reforms were to some extent stovepiped and insufficiently joint, but they did prepare the army for force projection. A key change intervened in July 1998 when the army separated combat and logistics into two functional commands: the combat command CFAT (eight joint brigades and five specialized brigades) and the logistical command CFLT (two support brigades).[35] CFAT was henceforth to concentrate exclusively on the training and preparation of expeditionary forces, while CFLT was to concentrate also exclusively on the logistics of projection.[36]

The CFAT is perhaps the most impressive evolution at the level of land forces command. It is a level 1 headquarters with four subordinate level 2 force headquarters (located in Besançon, Marseille, Nantes, and Limoges). In the course of 2003 it was decided to organize at CFAT Lille a rapid reaction corps headquarters (HQ CRR-FR) intended not only to be rapidly deployable but also to enable French command in combined operations. Approximately 20 percent of the CRR-FR staff is international, drawn from partner nations (members of either the EU or NATO), which is to say around 75 of its total staff of 415, which can be augmented to around 700. HQ CRR-FR gives France the ability to act as lead nation in a multinational operation involving up to 50,000 soldiers, which happens also to be the EU headline goal.

We should note also France's participation in the Eurocorps. Besides the uniquely binational Franco-German Brigade, France assigns to this corps one of the level 2 force headquarters (Marseille) with one armored brigade, one mechanized infantry brigade, and support units, if necessary. In 2003, the Eurocorps headquarters was certified as NATO high readiness land forces headquarters (HRF(L) HQ), one of six to attain this status.

We should note finally changes at the joint level of command in France that began in 2002. The joint level was created in the early 1990s following the Gulf War experience with the establishment of a joint general staff (EMIA) and a joint operational center (COIA).[37] EMIA was supposed to do "cold" planning and COIA "hot" planning. However, the "hot-cold" setup was too slow and too narrow, which became apparent notably in the context of Kosovo and Afghanistan.[38] Or, as one staff officer noted, the setup was a "mess."[39] In 2002, as part of the preparation for the 2002–8 program law and as part of the response to the promise of providing military strategic leadership capabilities to the European Union—a promise issued at the 2002 Nice summit—France reformed the joint command organization. Strategic planning now happens uniquely in the planning and operations command (CPCO), while operational planning and command is located in a joint force and training command (EMFEIA).[40] France is thus better equipped today to take command of large combined and joint expeditionary forces, although, as the officer noted with an understatement, it will not happen every day.

EFFECTS-BASED OPERATIONS

Effects-based operations (EBO) represent a paradigm change in the application of military force insofar as it seeks to apply force in order to obtain strategic effects and not a priori to destroy enemy forces. In the past, force destruction was the ultimate strategic effect in war; today, NCW and expeditionary forces enable a different conception of strategic effects. NATO in 2006 issued a Concept for Alliance Future Joint Operations (CAFJO) that will embed EBO in allied planning. NATO allies began already in 2001 to organize Multinational Experimentations (MNE) (de facto organized by the US Joint Forces Command) focused on jointness and effects-based planning, execution, and assessment in a highly complex environment. EBO raises many questions. How will it affect service jointness? How will combined operations move focus to problems of multinationality? And how can one strike a balance between the political and military problems involved?

France reacts to EBO ideas emanating mainly from an American context, but, like Britain, France does not simply import the new ideas. As an equivalent to Britain's wish to downgrade ambitions—national as well as allied—with the new heading Effects-Based Approach to Operations (EBAO), the French ambition is to promote a "synergy of effects." That much is clear from the third and final constitutive concept of the land forces' transformation blueprint.[41] Below we will first trace the contours of this synergy-approach, then assess political implications, and finally trace the interaction of military and political issues in the context of the MNEs.

Synergy of effects is most notably an attempt to anchor EBO at an operational level while remaining open to its obvious strategic importance. The constitutive concept thus underscores "French military culture" and its traditional emphasis on force plans that divide missions into "effects to be obtained at every level."[42] French military forces thus act within the context of one overall "major effect" (effet majeur) to be obtained. The introduction to the constitutive concept explicitly states that British-American EBO ideas have not been directly translated into this French concept but have merged with the "major effect" tradition to become the final "synergy of effects" approach. French military traditions and history are also emphasized early in the concept where the desire to create "synergy" is traced to both General Jomini and Marshall Foch.

This effort to distinguish the synergy approach may now turn out to be astute. EBO is in general marked by ambiguity between its operational and strategic implications, and its operational version most in vogue in the United States has fallen into disrepute. General Lance Smith, head of NATO's Transformation Command (ACT), noted in the fall of 2006 that "there are lots of things in the effects-based thinking world that are difficult to define." Tactically, it is a question of informing a commander of desired effects and letting him make military decisions on how to achieve them; strategically, you involve other instruments of power and other countries. At this higher, strategic level, "you get more confusion, but the concept is still the same."[43] There may be quite a lot of confusion at strategic levels, but there is even more at operational levels, concluded the successor of General Smith, General Mattis, in mid-2008.[44] General Mattis, head of the US Joint Forces Command and also ACT, has now banned the use of operational EBO. However, he maintains that strategic EBO is useful, and it is that version of EBO to which both NATO and France subscribe.

The French conception of synergy springs first of all from the desire to combine several types of forces to achieve a "global operational effect." To do so, and borrowing from EBO, it defines an end-state (*état final recherché*, EFR) which, in accordance with the "major effect" tradition, will be translated into action plans for different force levels. Synergy then results in a number of capability demands:

- real-time targeting and shooting;
- real-time target surveillance;
- dispersal of forces and concentration of effect;
- new command and control systems;
- precision munitions.

At this level "synergy of effects" is operational EBO and an outgrowth of the information technology inherent in NCW and the mobility inherent in expeditionary warfare—the two preceding constitutive concepts. It is deliberatively calibrated to fit these two other drivers of French Army transformation; it is not simply a new paradigmatic targeting doctrine grafted onto the army corpus. In working on these developments France is moving in the direction of creating theater command and control systems that will allow for integrated theater analysis, decision-making, and command. France is likely to continue down this path, even if the operational EBO ideal is in crisis.

This brings us to the more important strategic version of EBO, which follows from the combined and joint nature of most of today's wars. French land forces will have to interoperate with other military as well as diplomatic, judicial, economic, and humanitarian services, and all of them of multiple nationalities. This is what General Smith had in mind with the reference to EBO's strategic dimension. The best illustration of the challenge inherent herein comes from Afghanistan, where NATO since 2004 has been in command of the peacekeeping operation. NATO can run this operation only if its members can plan in common for it; generate required forces; fund it; and ultimate tie military success to civilian reconstruction and political development. The former issues—planning, force generation, and funding—define the so-called Norfolk agenda for NATO change and make up the essence of NATO's operational challenge in a new security environment.[45] The latter issue is EBO at its very broadest: achieving the military, political, economic synergy necessary for operational success. At this level, EBO for NATO entails the establishment of a permanent relationship with the European Union, a working relationship with the United Nations, and ad hoc arrangements involving concerned nations in specific operations.

The point here is that these two sets of issues—from force generation to EU-NATO relations—form part of one overriding political problem, when viewed from France. France is investing in military forces capable of participating in EBO-like coalition operations, but France, naturally, is not ready to witness politically that such coalitions promote diplomatic solutions inimical to French interests. The red lines for France concern the authority of the UN Security Council to authorize military operations; the autonomy of the European Union; and the status and influence of France itself. Put differently, France does not wish to see military operations launched without UN authorization; to see NATO become a framework for EU policy; or to experience the sidetracking of France in coalition operations. France has therefore been reluctant to adopt the Comprehensive Planning and Action (CPA: a kind of strategic EBO approach) debated by NATO allies since 2003: if NATO operations could thus involve a "comprehensive" framework, then NATO might begin setting the pace for the largely civilian EU, which in diplomatic parlance is referred to as "Berlin Plus in reverse."[46] CPA is a general policy based on a single case, Afghanistan, the French argue, and while it may work there it should not be turned into a general policy framework.[47]

Although there was no CPA consensus at NATO's "Afghanistan summit" in Riga, December 2006, NATO deputy ambassadors since managed to work out a CPA document enumerating the kind of pragmatic initiatives NATO could undertake in the spirit of "coordination." Returning to the above-mentioned MNEs organized by the USJFCOM, we should note that these took off in 2001 without the participation of France.[48] The purpose was to test various concepts for sharing information in coalition operations, such as force generation, technology requirements, and effects-based operations. France joined the third MNE in February 2004, participated also in the fourth MNE, February 2005, and is involved in the preparation of the fifth, set to run from mid-2008 to the spring of 2009. The 2004 MNE was geared notably to test NATO's new Response Force, and as a key NRF participant it was obvious for France to seek MNE participation. The 2005 MNE and the upcoming one focus more broadly on the many dimensions of coalition operations. Revealingly from an EBO perspective, France in 2005 took the lead in the MNE working group, "strategic context and conflict resolution," and in 2008–9 France will be in charge of "strategic interagency multinational planning." In other words, France intends to shape the strategic EBO debate.

France is worried not only about a "Berlin Plus in reverse" but also about other strategic EBO effects. The Kosovo War in 1999 involved notably a debate between allies on the extent to which targets should be subject of political debate, and, in conjunction herewith, whether bombing should be tactical or primarily strategic. France was among the allies insisting on strategic coordination.[49] While wars are no longer fought by "committee" but by coalitions, the issue remains. France has on some occasions protested against operations undertaken in Afghanistan because France, a coalition partner, was not informed and believed that the operations carried a large risk of undermining the larger mission—that is, of undermining the strategic effects. One such example was the Canadian-led Operation *Medusa* of the fall of 2006, in which some coalition partners, such as France, allegedly received too little information too late.[50]

Future events to be followed include the internal French organization of strategic EBO issues. Naturally, the real decision-maker in France is the president. However, interministerial coordination falls within the powers (defined by the constitution) of the Secretary General of National Defence (SGDN), who is attached to the prime minister's office. This will now change.

In 2007 Nicolas Sarkozy as candidate for the presidency announced the ambition "to create, under the President of the Republic, a National Security Council that will become the sole authority for analysis, debate and study of all matters relating to security and defence." It will substitute for current structures; have a permanent secretariat; and involve "government officials, heads of departments concerned and experts."[51] Once elected, in May 2007, President Sarkozy delivered on his promise. He first of all initiated a strategic reassessment that resulted in a June 2008 White Paper. Unlike its 1972 and 1994 predecessors, this White Paper does not merely deal with "defense" but "defense and national security." The title and the agenda of the White Paper Commission were thus infused by strategic EBO thinking. Moreover, the White Paper process has been directed by the SGDN, not the Ministry of Defense, which again reflects strategic EBO thinking.[52] Finally, Sarkozy will henceforth head a new Defense and National Security Council whose prerogatives will extend to any question and public service that touches on defense and security issues.[53]

The organizational and political implementation of strategic EBO will be interesting to follow. Equally interesting will be evolutions in the "synergy of effects" approach within the land forces. The "synergy of effects" approach will be affected one way or another by the US decision to downgrade operational EBO, but it is too early to tell how. It might be tempting for the French Army simply to fall back on its *effet majeur* tradition, but there is no natural affinity between this tradition and the EBAO approach under development in NATO where French allies bring their own individual traditions to the table. It will not be easy for NATO to define a simple and workable EBO approach, but we may surmise that nations such as France and Britain will play lead roles in finding a middle ground.

MILITARY CHANGE AND CHALLENGES IN FRANCE

It is apparent from this analysis that military transformation is a dominant theme in French defense planning. France is investing in NCW, force projection, and effects-based operation doctrines and approaches, and this as part of an effort to be at the forefront of military affairs.

Change really began in the early 1990s as a consequence of the Gulf War. Investments in new technologies followed, as did conceptual work to break out of the homeland habit, and the 1996 decision to professionalize the forces and aim for a 2015 model capped off the process. By the turn of the century, and in fact a few years into it, it became apparent that French forces had acquired many new technological and technical capabilities, but also that they lacked "clear concepts" to guide them, and that European-American cooperation was a prerequisite for success.[54] Transformation then became an issue.

Why did things happen this way? There is no question that the traditional

Gaullist preoccupation with national rank and the military's ability to bolster it accounts for some of the motivation of the Chirac reforms in 1996, just as the lessons of first the Gulf War and subsequently Bosnia are important factors: to be militarily relevant, France needed fundamental change. But why, then, did transformation become an issue only at the turn of the century? The international dimension is important: transformation crystallized as an agenda in the United States only by the late 1990s and really gained prominence in 2000 when Donald Rumsfeld became secretary of defense. Moreover, the French 1996 agenda for change was impressive in scope and therefore consumed most of the organizational and budgetary energy of the armed forces. Finally, the onset of political cohabitation in 1997 entailed a political downgrading of military reform.

Yet transformation did eventually take off. There are several causes for this. The transformation of US forces is a constant background factor that must not be overlooked, and that takes on particular importance when French forces are operationally in touch with them. It began with the Gulf War and continued with the Kosovo War. The 2001–2 war to topple the Taliban regime reinforced the point: from Kosovo to Afghanistan, the transatlantic technological gap has deepened and there is a real risk, concluded a major French daily newspaper, that American hardware will become too sophisticated to be compatible with that of the Europeans.[55]

The concern with military prowess is deeply rooted in French political culture and clearly affects the military institution itself as well. There is no easy way to demonstrate this cultural effect, though, apart from pointing to the motivations of the Chirac reforms, Sarkozy's security sector reform, and the various testimonies of military officers. It seems incontestable, though, that French political culture required multiple shocks to embrace transformation as an agenda, all linked to the operational loss of political influence, and we might venture that it is only Sarkozy—elected in 2007—who at a political level has embraced transformation as an agenda.

Going back to the late 1990s we know that the DGA became a driver of transformation. Why did the DGA take on this role? Part of the reason was budgetary, given the prospect in 1997–98 of a 30 percent cut. Part of the reason has to do with the traditionally close relationship between the DGA and the French defense industry. The profitability of that industry depends on the ability of the French forces to demand and use weapons and weapons systems that are globally competitive, which is to say that French forces must be transformed for the French industry to be competitive. Several background interviews underscored the importance of this DGA-industrial connection, and more evidence can be gauged from the efforts France invests in the EU's defense agency (EDA). Very few defense actors in France would disagree with these sections of the EDA's October 2006 Long-Term Vision:[56]

> If Europe is to preserve a broadly based and globally competitive [defense technological and industrial base] (which means competitive with the US, and, increasingly, producers in the Far East) it must take to heart the facts that the US is outspending Europe six to one in defence R&D; that it devotes some 35% of its defence expenditure to investment (from a budget more than twice as large as that of the Europeans combined), as against the European level of about 20%; and that it is increasingly dominant in global export markets. . . . Un-arrested, the trends points towards a steady contraction of the European defence industry into niche producers working increasingly for US primes. A combination of countermeasures is necessary.

The DGA role directs us to two challenges for France. First, its European partners must be convinced that a fully fledged industrial policy is in order, and also that such a policy might be part and parcel of a larger policy of organizing a defense core consisting of the six greatest EU powers—a G6 in European defense.[57] Secondly, France must organize its defense industry at home. The two competing architectures for the future land force (SCF) are a case in point. It may be advantageous for the public purse to entice competition between the defense industries, but it raises two related questions. How can these industries be kept in competition if all France really needs is one system? How can national consolidation be made compatible with the transnational exigencies of American industrial dominance and an emerging European industrial policy? Will access to the US market—such as the still likely though contested $40 billion contract in refueling aircraft awarded by the Pentagon to EADS-Northrop-Grumman—affect the French preference for a European defense market vis-à-vis a transatlantic market?

One critical actor in sorting out these issues will be the Chief of Staff (chef de l'état major des armées CEMA). We noted earlier that the CEMA entered the transformation process at a relatively late stage, in the course of 2005. Interestingly, the CEMA at the time, General Henri Bentegeat, was in November 2006 appointed head of the EU's Military Staff and moved to the center of the European defense debate and decision-making process. Incidentally, Bentegeat's successor as CEMA was another army general, Jean-Louis Georgelin.

The CEMA institution has been reinforced in various ways to strengthen central control, which can be seen in the decree of May 2005 defining CEMA powers, replacing the previous decree dating back to 1982.[58] The gist of the changes is threefold:

- CEMA no longer "assembles" capability proposals of the DGA and the services; he is responsible for their "coherence" and is authorized to make decisions before submitting proposals to the minister.[59]
- CEMA has gained authority in the day-to-day activities of the services, which is to say the activities unrelated to operations (the "organic" functions, in French). The service chiefs, previously in charge, now answer to the CEMA also in these matters (CEMA always retained operational authority).[60]
- Finally, CEMA is strengthened in all joint matters. In 1982, these were hardly developed; today, CEMA has authority over all joint institutions and is responsible for joint doctrines.[61]

This formal strengthening of CEMA in the spring of 2005 was accompanied by practical measures as well. One was the final and overarching transformation blueprint for all French forces—including the land forces, of course, but not exclusively so—that General Bentegeat approved in April 2005: the aforementioned General Policy on Transformation.[62] The approach adopted here is distinctively bottom-up: CEMA is to "invigorate and federate various initiatives in order to enhance the efficacy of military operations." The services retain influence and initiative, clearly, but CEMA is there to ensure coherence and collective drive. CEMA now also benefits from the creation of a Joint Center for Concept, Doctrine, and Experimentation (CICDE).[63] The CICDE began operating in 2004, although it was formally established by decree only in April 2005 in parallel to the strengthening of the CEMA. CEMA faces three distinct challenges as it pursues transformation French-style.

First of all, the CICDE is a young and small organizational body that must teach old and big services new ways of thinking. There is no question that CICDE is there to advance new thinking. One CICDE official noted that CICDE is intended to be a distinctively rapid and flexible mechanism for testing concepts and writing updated joint doctrines that can act as "instruments of transformation."[64] There is some way to go, though: a full set of relevant doctrines must be worked out; they must then become embedded at the joint level; and they must become part and parcel of the services that already develop regular service doctrines. The service most in need of doctrine is the land force, because it is the biggest and most complex organization in terms of humans, machines, and operating environment. The question thus is whether land force commanders will read CICDE doctrines, reflect on them, and finally act on them and disperse their lessons to the individual soldier. This question in turn depends on the working relationship between the army's doctrinal center (CDEF) and CICDE. CDEF is bigger and more experienced and likely to resist its subjugation to dictates of "jointness." CEMA, to move transformation forward, must invest part of his prestige in CICDE, as well as time and effort to clear up the CICDE-CDEF relationship.

Secondly, CEMA must give direction to the R&D and procurement process, which involves strong vested interests—notably the defense industry and each of the three services, with the DGA officials situated somewhere in between. The services have platforms and programs to defend (such as Scorpion); the DGA has an architecture to defend; and the industry has projects to defend. One problem is that the DGA does not really work collectively with the services and the industry: it works instead first with one and then the other. A DGA force architecture (for example, Scorpion) may thus have rival interests built into it. CEMA must notably balance these interests and support an architecture that ultimately reinforces the culture and capacity of the service tasked with a particular job. One army official expressed a hope for an overall policy of "subsidiarity,"[65] in the hope of protecting the service, but the fact of the matter remains that it is too early to tell how CEMA will tip the balances.

Finally, there is a challenge of creating a coherent culture for the armed forces as a whole. We saw earlier that the land forces were worried that modern expeditionary warfare might rob the individual soldier of emotional attachment to the fighting group and thus lead to a "cultural rupture." The point applies to the full set of forces, although the land forces are a particular case given the role of human beings in their complex environments. The collective challenge for CEMA and the three services concerns the basic sense of purpose. France now pursues transformation—albeit bottom-up as opposed to the perceived American top-down approach—because it wants to be at the military forefront. But France also tends to deny the ambition to "fight wars" because France is out to "control violence" and do "peacekeeping." We witness this benevolent ambition in army doctrine as well as political speech,[66] notably that part of it related to the EU. The question is whether such ambitions can be said to inhibit the doctrines and force structures that are needed to realize transformation because transformation must involve a willingness to fight and win wars.[67]

The record provides no clear answers in this debate. While it appears true that today's complex operations provide few easy lessons and threaten to enmesh military forces in settings so complex (by virtue of interservice, interministerial, interallied, and interinstitutional relations) that their savoir-faire could be undermined, it is equally true that there are few, if any, real wars to fight and win. Transformation, it seems, is bound to be tied up with complex operations, and this is in fact a starting point for French defense reforms. What seems certain is that transformation, if it is to happen in France, must move from overall planning to operations and operational impact. Overall planning is happening, qua transformed units such as the 2nd and 6th Brigades and their deployments, but the operational impact is less developed. France is facing a potential division between the engineers and force planners whose task it is to design tomorrow's technology and the army soldiers who today must engage in fights that in terms of complexity and asymmetry resemble past and not future warfare.

France deployed a force of around 700 soldiers to eastern Afghanistan in the course of 2008 following a commitment to that effect at NATO's April 2008 Bucharest summit. Fighting in eastern and southern Afghanistan has become the real engine of transformation for the allied forces operating in those regions, and in this sense France is right: real transformation must take place bottom-up. The trouble for France is that its commitment to fighting in these regions is tenuous at best. President Sarkozy seems motivated by the bigger issue of NATO-EU reconciliation and France's opportunities for leadership: to claim it, France must have boots on the most difficult Afghan ground. The French military appears to see no real merits in this affair. No effort has been made to instill in them the sense that this is their war. Moreover, the president who commanded them into the fighting also commands them—via the June 2008 White Paper—to undertake new great reforms at home, which will be costly in terms of time and money and, notably, organizational tradition and history. Civil-military relations are strained as a result. Afghanistan is therefore not likely to become France's engine of transformation in the near future, and in this sense transformation remains a blueprint whose realization is uncertain.

CONCLUSIONS

France has engaged the process of military transformation in all important dimensions—from network-centric warfare over expeditionary forces to effects-based operations. The process began around the turn of the century and picked up pace in 2003–5. Today, there is an overall transformation blueprint for the armed forces and an elaborate blueprint for the service most in the need of transformation—given its size and territorial legacy—the army. These blueprints tie in with the 2008 White Paper, the long-term procurement programs that will follow, and a political desire to think far ahead.

The French trajectory is inspired by that traced by US forces, though national particularities are visible. France does not to the same extent seek to create systems-of-systems, given its focus on network enabling as opposed to network-centric warfare. France has yet to define its future land force, unlike the United States, in the belief that it is too early to tell given the technological state of the art as well as political and economic conditions. France is definitely developing expeditionary capabilities but the process is framed partly by a political and doctrinal focus on conflict resolution and crisis management, as opposed to war, partly by the ambition to become a European lead-nation, which results in a certain tailoring of army and joint capabilities within France. Still, the lead-nation ambition is considerable and thus a cause of considerable investments and capabilities. France is also pursuing the idea that military operations should not seek the destruction of enemy forces as much as generate desired strategic effects of another kind—EBO. France has developed a "synergy of effects" approach drawing on its traditional approach to operations, amounting to an evolutionary approach. Still, France is heavily involved in the transatlantic search for a common EBO framework, although perhaps predominantly at strategic levels. Strategic EBO and thus strategic leadership nourish longstanding French ambitions to be part of the inner Atlantic circle dominated by Anglo-Saxon countries.

Two challenges stand out. One is the past absence of a strong central player in the military process and the positioning of CEMA as a new strong player. His ability to influence a large and complex process is varied, but one of his options is to appeal to the traditional French political concern with the military's ability to strengthen French foreign policy. Whether parliamentarians and policymakers will give support military transformation with coherent budgets, industrial and alliance policies is another question. Another challenge concerns the translation of transformation blueprints into doctrines that reflect an operational view of how to fight wars differently. This is notably a challenge for the army which must fight its wars on the ground where complexity reigns. The support granted in Paris to transformation blueprints will tie in with the operational experiences of the army to decide the fate of transformation *à la française*.

5 The Rocky Road to Networked and Effects-Based Expeditionary Forces: Military Transformation in the Bundeswehr

Heiko Borchert

Today the Bundeswehr is in transition. The days of the old Bundeswehr fo-cusing on territorial defense are gone. The new Bundeswehr that operates in international missions with allied partners on three different continents has become reality. Transformation has started to take ground. Excellent concept papers flesh out the vision of an agile force that makes optimal use of its re-sources by equally balancing people, organization, and technology, leverages the benefits of superior approaches toward knowledge creation and diffusion in an effects-based environment, and interacts closely with other instruments of power in a networked approach to security. In doing so, Network-Enabled Capabilities (NEC) have been identified as one of the main pillars of German defense transformation, and the development of an overarching Role-Based Common Operational Picture (RobCOP) is an important catalyst to providing joint situational awareness and joint situational understanding. Based on these ideas, the German Army, Air Force, and Navy, the Joint Support Service, and the Medical Service will embark on specific transformation avenues that help improve the operational effectiveness of the Bundeswehr as a fully integrated and interoperable "system of systems."

However, the reality of Bundeswehr transformation, which was officially kicked off in 2004 with the new *Concept of the Bundeswehr*, does not yet fully live up to this vision. Implementation deficits, most importantly with regard to mis-sion-oriented requirements, the loss of powerful promoters, the lack of systematic change management that leads to shortfalls in human resources and communi-cation policies, as well as funding restraints, make it difficult to realize transfor-mation as foreseen. As a result, change has not yet captured the Bundeswehr as a whole. Rather, it has led to different islands of progress and innovation. This is not to say that Bundeswehr transformation has failed or has come to a standstill. But it is fair to concede that the cultural and structural premises for transforming the armed forces in Germany are particularly more difficult than in other countries. In a country in which "friendly public disinterest" and a "cautious public distance vis-à-vis military power" dominate,[1] it requires extra political will to accept the Bundeswehr as an instrument of power and to provide it with the necessary financial, personnel, and material resources needed to accomplish its tasks. Right now, this extra political will is in short supply.

This paper shows that the Bundeswehr is fully engaged in a demanding a transformation process with a distinctly German footprint. Bundeswehr transformation is understood as a comprehensive undertaking. It takes place within the broader framework of networked security that emphasizes the need to balance military and nonmilitary capabilities to tackle today's and tomorrow's security challenges. The first section outlines the German understanding of defense transformation. Then the study looks at the implementation of transformation. After outlining the change in favor of international operations, the four key transformation strands—NEC, RobCOP, Effects-Based Approach to Operations (EBAO), and Concept Development and Experimentation (CD&E)—will be discussed in detail. This is followed by an initial transformation scorecard highlighting accomplishments and room for improvement. The final section explains the Bundeswehr transformation process by help of the theoretical literature on transformation summarized in the introductory chapter by Theo Farrell and Terry Terriff.

TRANSFORMATION APPROACH

The German understanding of transformation has a slightly different footprint than approaches followed by other countries. Germany shares with its EU and NATO allies the overall goal of sustainably improving the operational effectiveness of the Bundeswehr through transformation.[2] This process takes place against the background of a changing strategic landscape that requires security institutions that provide enough organizational flexibility to cope with these changes. Transformation in the German understanding is seen as "the proactive shaping of a continuous process of adapting" the Bundeswehr and the German security sector.[3] Therefore Germany puts more emphasis on the broader strategic framework within which the continuous development of the Bundeswehr takes place.

This overall framework is called networked security, which is comparable to NATO's Comprehensive Approach.[4] Networked security is an all-encompassing philosophy that builds on the comprehensive understanding of security. It aims at providing superior security-relevant effects by balancing all relevant instruments of power in a concerted effort. Networked security thus advances the idea of systematically interlocking all relevant security sector actors, levels of decision-making and implementation (from the international level within NATO, the EU, and the United Nations to local levels of interaction), security instruments (diplomacy, information, military and law enforcement, economy), and tasks to be accomplished (conflict prevention, conflict management, and postconflict stabilization).[5]

The need for transformation of the Bundeswehr within the framework of networked security basically results from new security challenges and the blurring of distinctions between traditional security concepts such as domestic versus international security, military versus nonmilitary approaches, or public versus private responsibilities. In order to cope with the new environment the Bundeswehr has improved its ability for organizational learning and structural flexibility. Technological innovation, in particular in the field of information and communication technology, can assist this process. This, however, demands more than procurement modernization. What is needed is a new informational infrastructure that helps link sensors, decision-makers, and effectors. By harnessing the power of information and adopting new concepts, the Bundeswehr—together with its civil partner ministries—aims at achieving effects superiority. This will affect organization/service, personnel, training/exercises/missions, material/equipment, and concepts and methods/procedures, which have been identified as the six areas of work to make transformation happen.[6]

In order to implement defense transformation a dedicated structure was established. Overall responsibility for planning rests with the Chief of Defense, Bundeswehr. The Deputy Chief of Defense, Bundeswehr is personally in charge of all transformation activities of the Bundeswehr. A Coordination Group Transformation led by the Planning Division in the Armed Forces Staff was established to bring together all relevant elements from the military and civil side of the Ministry of Defense. In addition, the Bundeswehr Transformation Center was established as the key enabler between the Armed Forces Staff and the Service Commands.[7] In order to provide guidance for transformation, the Coordination Group Transformation has produced a so-called Master Plan

Transformation. The Master Plan serves as a reporting document to assess what has been accomplished in the past, what is existing at the moment, and what might be interesting in the near future.[8] But, as will be discussed below, current structural differences between the different services of the Bundeswehr make it difficult to use the document as a guide for mission-oriented implementation of the Bundeswehr transformation.

Compared with other case studies in this volume, Germany's transformation process is very much concept driven. Other countries follow a more pragmatic and hands-on transformation approach that aims for quick wins. The Bundeswehr, like the German public sector in general, by way of history and cultural heritage, has a tradition for serious conceptual work and sound deliberation before actions are taken. This inclination dates back to Max Weber and his analysis of the inner workings of bureaucracies. The bureaucratic culture in Germany also explains why it is so important to get new conceptual ideas into the paper trail before things get done. This is noteworthy because the ideas of transformation and NEC entered the new Bundeswehr concept papers relatively late.

In 2002 the then Bundeswehr Center for Analyses and Studies (now the Bundeswehr Transformation Center) published a landmark foresight study comparable to the Global Strategic Trends Programme by the UK Development, Concepts and Doctrine Center.[9] This study clearly highlighted the need for continuous change within institutions as a key feature of the new security environment, argued in favor of a new capabilities-based planning approach, advanced ideas for closer cooperation between the Bundeswehr and the defense and security industry, and provided the basis for what have later become the six generic core capabilities of the Bundeswehr. While some of these ideas found their way into the Defense Policy Guidelines that were presented by then Defense Minister Peter Struck in 2003, there was no mention of transformation or NEC in his guidelines. The same almost also happened with the *Concept of the Bundeswehr*, published in 2004. Only six weeks prior to its publication, this document did not include any reference to the new ideas.[10] This explains why the *Concept of the Bundeswehr* provides only conceptual "flash lights" rather than a coherent vision of defense transformation.

TRANSFORMATION CONCEPTS AND AVENUES

The Bundeswehr transformation process is implemented along four intertwined avenues: EBAO aims at improving operational planning, while contributing to a better balance between all instruments of power. Achieving mission effectiveness at the tactical level, in particular via NEC, is the core of military transformation. It is built upon the provision of a Role-Based Common Operational Picture as the key joint undertaking to advance situational awareness and situational understanding. Finally, Concept Development and Experimentation (CD&E) is seen as the central methodology or "workbench" to test new as well as alternative concepts. These four strands contribute toward improving the Bundeswehr's operational effectiveness, in particular for expeditionary operations.

Expeditionary Operations

At the end of 2008, the German Bundeswehr participated in several international operations, with around 6,600 soldiers. The year before, around 30,000 soldiers out of the new target size of 250,000 soldiers were bound by operations, preparing for or returning from operations.[11] Since 1992 Germany has spent €10.5 billion on international deployments of the Bundeswehr.[12] The development that has taken place since the end of the Cold War was far from obvious at the beginning.[13]

The Bundeswehr inherited many different Cold War legacies that proved burdensome for the first years of Germany's reorientation. Two of them should be highlighted in particular. First, the reunification of two different German states brought with it the need to integrate the armed forces of the former German Democratic Republic into the Bundeswehr and the redeployment of 500,000 Soviet soldiers stationed in Germany at that time. This absorbed a lot of attention, resources, and time during the crucial years of strategic reorientation immediately after 1990. Second, for historical reasons the German constitutional law puts very strict limits on the use of force for domestic and international political purposes. This lasted until 1994, when the German Constitutional Court set the ground for Bundeswehr participation in international combat operations under the mandate of international organizations.[14]

Since then the political decision about "going expeditionary" has been directly tied to the effective right to veto of the German Bundestag. In light of Germany's history and its self-understanding as a civil power this provision can be understood as a barrier that should help prevent Germany from slipping into "military adventures." In practice, however, doubts have been raised with regard to the compatibility of Germany's parliamentary processes and the underlying political will for military deployments abroad on the one hand, and evolving international requirements for swift and decisive decision-making on the other. In addition, only the deployment of the Bundeswehr requires a mandate of the Bundestag. Sending police forces abroad or dispatching Germany's development assistance agency, GTZ, to international zones of crisis—both on a voluntary basis—takes place without such an act. This leads to a situation in which the use of the armed forces is under constant political observation whereas other areas of Germany's foreign and security policy almost evade the political radar screen. This must be reconsidered under the newly formulated goal of networked security.

During the 1990s Germany accepted the geostrategic changes imposed by the end of the Cold War but only slightly changed the Bundeswehr. Under Defense Minister Volker Rühe (Conservative, 1992–98), the first international deployments of the Bundeswehr (for example, Cambodia, Somalia) were undertaken. His successor, Rudolf Scharping (Social Democrat, 1998–2002), launched a modernization program aimed at leveraging the benefits of outsourcing and public-private partnerships. He also introduced corporate management instruments in the Ministry of Defense. The overall conceptual reorientation of the Bundeswehr began with Peter Struck (Social Democrat, 2002–5). In 2003 he published the defense policy guidelines. These guidelines acknowledged that there is no longer a territorial threat to German security. Rather, and in line with the European Security Strategy that was adopted in the same year, the new security environment is characterized by an explosive mixture of new risks such as international terrorism, the proliferation of weapons of mass destruction, regional conflicts and crises, failed states, organized crime, pandemics, migration, and cyber threats.

The 2003 defense policy guidelines stipulated—and this was new—that the requirements of international operations determine the mission, tasks, and capabilities of the Bundeswehr. Germany's security, Struck used to say, is also defended at the Hindu Kush in Afghanistan. In line with this statement, the Bundeswehr was expected to fully operate in a multinational environment. Germany will act nationally on its own only in evacuation operations. For all other international contingencies, the Bundeswehr will operate hand in hand with its partners.[15] In order to prepare for the growing international role of the Bundeswehr, Struck's defense policy guidelines introduced six capability categories that are intertwined: command and control, intelligence and reconnaissance, mobility, effective engagement,[16] support and sustainability, and survivability and protection. Since 2003 these categories have become the conceptual cornerstones for Bundeswehr development, harmonizing operational needs and armament procurement activities.

Many of the international operations assumed over the past ten years have brought fundamental changes. In 1999 Germany for the first time supported a military intervention out of area by providing support for Operation *Allied Force* in Kosovo. Only two years later the German Bundestag gave the green light for deploying Special Operations Forces to Afghanistan as part of Operation *Enduring Freedom* (OEF). In 2003 the German ministries of Defense, Foreign Affairs, Interior and Economic Development, and Cooperation have jointly begun to field so-called Provincial Reconstruction Teams in the northern part of Afghanistan. Blending civilian and military capabilities in a new interagency approach can be seen as a first attempt to implement the idea of networked security. In 2006 the United Nations mission in Lebanon brought German forces close to Israel, where the German contribution to provide stability with maritime forces was explicitly welcomed. Finally, in December 2008 the German Parliament decided to support the first European Union naval operation, *ATALANTA*, to fight international piracy off the coast of Somalia, with a contingent of up to 1,400 soldiers and a frigate.

The process of going international was accompanied by substantial structural changes. Most important, the Bundeswehr set up special elements to command and control international operations. At the ministerial level a new Operations Staff (*Einsatzführungsstab*) was established in mid-2008.[17] Its main function will be the coordination of all mission-related military and civil tasks. The core of the new Operations Staff was formed out of the existing "Bundeswehr Operations" Division in the Armed Forces Staff. The Operational Headquarters (*Einsatzführungskommando*, OHQ), which was established in 2001 in Potsdam, serves as the core center. In addition, the Response Forces Operations Command in Ulm (*Kommando Operative Führung Eingreifkräfte*) is foreseen as a deployable force headquarters, in particular for EU-led operations. It took over OHQ functionality from the *Einsatzführungskommando* in 2008. In Kalkar the air force established a deployable Air Component Command and an interagency Air Defense Command (*Nationales Lage- und Führungszentrum für Sicherheit im Luftraum*), which brings together experts from the air force, the federal police, and the national air traffic control agency *Deutsche Flugsicherung* (DFS).[18] Furthermore, the Bundeswehr has also set up a Special Forces Command and reorganized special forces capabilities under the Special Operations Division, which provides adequate forces for use in today's asymmetric threat environment.

In addition, the Bundeswehr has become one of the few armed forces distinguishing structurally between three different sets of forces: response forces (35,000 soldiers) are aimed primarily at high-intensity tasks that require quick response and robust engagement in theater. Response forces will be fully network-enabled. Stabilization forces (70,000 soldiers) are used for low- and medium-intensity operations within the broad spectrum of peace stabilization. Stabilization forces must be capable of interacting with response forces. Support forces (147,500 soldiers) provide assistance at home and abroad.[19] Finally, joint intelligence collection and reconnaissance, logistics, and command and control (C2) support have been centralized for better performance in the new Joint Support Service, established in 2000.[20]

Being on international missions has become the rule for the Bundeswehr—a situation that puts the forces under heavy strain. The current level of ambition is set for five parallel stabilization missions in different theaters of operation with up to 14,000 soldiers at a given time.[21] This, however, is hardly achievable, not least because of unsolved manning and equipment problems. Urgent Operational Requirements (UOR) have drastically increased. They are not only absorbing the time of planners,[22] they also require more and more money. Over the past four years around €2 billion (or €500 million per year) have been spent on UOR. This is a significant amount, as the Bundeswehr has an annual defense procurement budget of only around €4 billion.[23] Therefore, it is hardly surprising that the Bundeswehr suffers from substantial capability gaps in areas most critical for international operations. Protection is one of the areas where existing shortfalls have led to heated political debates. For example, the Bundeswehr's much sought-after capability for combat search and rescue will not be ready as planned.[24] Further key capability shortfalls include the lack of the Bundeswehr's own strategic airlift, global reconnaissance, high-performing and interoperable C2 systems, initial air-defense capability against missiles, and the capability to locate precisely and engage a specific target.[25] As a consequence, structural disabilities that result from growing political commitments for international engagements on the one hand, and the lack of funds and political will to change the Bundeswehr according to these new operational requirements on the other, threaten the effectiveness of Germany's armed forces.[26]

Network-Enabled Capabilities

Achieving Network-Enabled Capabilities—called *Vernetzte Operationsführung* (NetOpFü) in German—is the main inroad to transforming the Bundeswehr at the tactical level. According to the 2004 *Concept of the Bundeswehr,* NetOpFü is based on the value chain of reconnaissance, command and control, and effective engagement. NetOpFü relies on a joint, multilevel, interoperable information and communications network that brings together all relevant sensors, effectors, operational units, planners, and decision-makers.[27]

Based on this general idea the Bundeswehr published a NetOpFü concept paper in 2006.[28] The paper shows that German thinking is influenced by US ideas, but is more comprehensive in scope. For instance, the German paper makes explicit reference to the four dimensions of NEC (social, cognitive, information, physical) introduced by the US Office of Force Transformation and to the US/British thinking on the NEC value chain comprising the three levels of information, decision, and effects superiority.[29] Against this background the paper identifies common situational awareness and common situational understanding as the key prerequisites of joint operations. In this respect, the Bundeswehr puts a premium on establishing a RobCOP. As will be outlined below, the RobCOP is an important catalyst for force transformation in Germany.

The paper also outlines an incremental path to achieving NetOpFü that covers three stages. First comes the improvement of joint situational understanding, inter alia, by establishing a RobCOP, integrating the above-mentioned network incorporating C2, reconnaissance, and effective engagement, as well as tying mobile, precision stand-off weapons systems of response forces into the network. By 2012,[30] limited initial technical preconditions and capabilities for the application of NetOpFü principles shall be demonstrated. The second stage aims at improving the quality of planning and decision-making processes by an advanced RobCOP, decision support systems, and adaptations of the C2 structure. The Bundeswehr's full NetOpFü capability will be reached once the network and the respective IT architecture that allows for collaborative interaction among all people and organizations involved also covers the lowest tactical level. To that end, reach-back between domestic units and units abroad will also have to be fully established. Fully operational NetOpFü capability shall be accomplished in 2020.[31]

The concept paper then continues to identify NetOpFü as a capability multiplier and identifies its added value to each of the six capability areas of the Bundeswehr. The paper concludes by outlining the need for action in each of the six areas of transformation. This is where the German approach goes beyond other nations' NEC concepts by highlighting what must be done in order to make sure that people, organization, and technology can be developed hand in hand. Finally, the Armed Forces Staff accompanies these processes by a joint NetOpFü working group that coordinates all relevant activities across all services as well as the civilian side of the Ministry of Defense. The IT Office of the Bundeswehr is in overall charge of technical implementation.[32]

While the path toward NetOpFü has been perfectly captured in different concept papers, implementation seems to be more problematic. First there is a lack of funding. This means that a homogenous multilevel C2 network can be achieved only in the mid and long term. Tangible NetOpFü capabilities will not become available until well into the next decade.[33] In addition, the Bundeswehr lacks an overall mission-oriented definition of requirements and an implementation-oriented road map for NetOpFü.[34] At the time of writing it is therefore difficult to say where the Bundeswehr stands right now and which units should achieve what goals at which point in time.[35] Finally, the current structure of the Bundeswehr seems to be a significant obstacle for joint operations. Initial assessments reports for the *Common Shield* Experiment, which was conducted in summer 2008 to evaluate the feasibility of NetOpFü, pointed out that there are significant differences in the way the army, the navy, and the air force interpreted NetOpFü principles as well as other military capabilities. From a conceptual point of view the services are expected to cooperate seamlessly as response, stabilization, and support forces, but in reality they continue to communicate, act, and operate differently. Among other things this outcome results from service-specific legacies.[36]

In order to tackle these problems it was suggested to push for the definition of specific requirements within NATO and the EU with regard to NetOpFü in general and the RobCOP in particular. This could have served as a hook up for transformation by designating those units for transformation-relevant tasks earmarked for the NATO Response Force (NRF) or EU Battle Groups (EUBG). That, however, has not yet happened. Therefore the NRF or EUBG do not serve as active drivers for the Bundeswehr transformation.[37] As a consequence there is a certain vagueness with regard to the milestones for NetOpFü mentioned above. Despite this problem, different working and coordination groups have been set up to make sure that developments in different areas are harmonized. In the Coordination Group Transformation, representatives of all relevant units make pledges for future achievements. Responsibility to the Chief of Defense is a mechanism to advance compliance, and there seems to be a certain "peer pressure" as everybody knows about each other's commitments.[38]

Furthermore it should be noted that NetOpFü does not belong to the six generic capability categories of the Bundeswehr. This explains why there is a certain lack of "organizational attention" attributed to this capability. As a consequence there are plans to provide different military units with NetOpFü-relevant capabilities. Rather than focusing on the establishment of dedicated experimentation units, the idea is to prepare different but regular units for NetOpFü tasks. In doing so, there would be a specific focus on covering the whole capability chain, for example to provide time-sensitive targeting or joint fire support. In providing the respective capabilities, new reference units would work on all different aspects of the DOTLMPFI spectrum.[39] In doing so, reference units would work hand in hand with the defense procurement division, defense planners, and the industry by way of spiral development.[40] Whereas the Armed Forces Staff seems to push this idea, others are more critical, arguing that there is not enough money for fully fledged jointness. Therefore reference units would probably be service-oriented, which could pose problems for jointness.[41]

Role-Based Common Operational Picture

For the Bundeswehr the RobCOP is one of the important catalysts to achieve NetOpFü. According to the relevant concept paper the RobCOP is a key instrument to fully leverage the potential of the information-based NetOpFü value chain. Based on the joint use of information gathered from different sources, the RobCOP contributes toward improved individual and joint situational understanding and thus facilitates and advances collaboration among all actors involved in NetOpFü-based operations. The RobCOP should display intentions and tasks of superior command levels. It should also provide a comprehensive assessment of relevant information regarding the situation of the Bundeswehr in a theater of operation and the situation of the enemy or conflict parties to be met there. Finally, the RobCOP should also assess the mission-relevant environment, including all relevant actors on the ground (for example, nongovernmental organizations).[42]

An analysis of the status quo prior to conducting the first RobCOP experiment made it clear that existing operational pictures available in all Bundeswehr services were insufficient for combined and joint operations. This was a technical (lack of interoperability) and a conceptual problem (lack of standards to generate operational pictures). In addition not all relevant information was available. In particular there was no systematic information fusion with nonmilitary actors. Finally, the IT system of the Bundeswehr included only limited capabilities to display all the functionalities of a RobCOP that evolves dynamically. This led to the conclusion that a RobCOP must be interpreted as a capability that has yet to be established.[43]

Testing the RobCOP concept was the main purpose of the *Common Umbrella* exercise conducted in 2006. The scenario assumed a stabilization environment in which air-to-ground attacks should be avoided by way of establishing a concerted defense effort based on the RobCOP. Under the coordination of the Bundeswehr Transformation Center the army, air force, and navy took part together with the industry and other organizations of the Ministry of Defense. Information from all services was fused in order to jointly coordinate sensors and effectors for countermeasures.[44] The experiment underlined that a RobCOP can be established. Since then the provision of a RobCOP is a key issue addressed in every NetOpFü CD&E experiment of the Bundeswehr. Two main lessons have been learned: there is a need to analyze how a service-oriented architecture for NetOpFü can be designed that helps avoid the risks of micromanagement and the blurring of different command levels (the "long screwdriver" phenomenon). In addition, further experiments are needed to determine in more detail the needs of role-based information gathering and diffusion and to define the IT architecture commensurate with those needs.[45]

Currently, different activities are underway to advance the idea of a RobCOP. There is, for example, work on a RobCOP for the artillery C2 system of the army and for peace-building operations at the Army Training Center in Wildflecken. The problem, however, is that because of the lack of specific RobCOP requirements, it is difficult to assess progress and to coordinate tasks in light of a specific future accomplishment.[46]

Effects-Based Approach to Operations

Germany is a strong supporter of EBAO. From the German point of view, EBAO is directly linked to the new paradigm of networked security. At least in part it seems feasible to argue that the readiness of the Bundeswehr to embrace EBAO can be attributed to the German political and strategic culture. The traditional preference of diplomatic and economic instruments of power and the emphasis on a comprehensive understanding of security, which has been advocated in Germany since the end of the Cold War, provided a good basis for EBAO in principle.

As outlined above, the new defense white paper's most important contribution is the focus on networked security.[47] The defense white paper makes the argument that a new quality of interagency interaction is needed in order to integrate economic, diplomatic, military, and civilian instruments of power into concerted action. This visionary approach could serve as a basis to advance the European Security Strategy and is fully in line with NATO's EBAO approach. Therefore networked security opens many fruitful avenues for international cooperation. However, the innovative character of this approach was hardly noticed at home, since the publication of the defense white paper was overshadowed by a scandal. One day before the paper's publication, a leading German tabloid paper published pictures of German soldiers in Afghanistan showing off skulls like a battle trophy. The next day the scandal had caught media attention, and politicians reacted in disgust. The allegations, however, turned out to be insubstantial.[48] In retrospect it could be argued that this scandal was a deliberate information operation against the new white paper in order to damage public reception of the paper.

This problem notwithstanding, the Bundeswehr has not yet adopted a unified approach to EBAO. Given the focus on networked security, this is surprising. At the time of writing, an agreed national German position on EBAO was underway. From what has become known so far, the Bundeswehr seems to interpret EBAO as a dual-purpose concept: on the one hand it helps improve operational planning by providing a more holistic assessment of the situation on the ground and the various ways to achieve a desired political end-state. On the other, it could help advance interagency interaction that is indispensable to blend military and nonmilitary capabilities to the benefit of integrated security approaches.

Initial problems in defining Germany's position on EBAO seem to have resulted from disagreement between two key divisions in the Armed Forces Staff. The division responsible for Bundeswehr operations interpreted EBAO as a future concept, whereas the Division for Bundeswehr Planning saw it as an operational concept.[49] According to an interim solution the Division for Bundeswehr Operations in the Armed Forces Staff was tasked to be in charge of EBAO. Because of the establishment of the new Operations Staff, which was described above, the Division for Bundeswehr Operations was hollowed out. As a consequence an organizational focus point for EBAO is missing.[50]

It is said that the explicit formulation of a German position on EBAO was triggered in particular by two developments: First, the Alliance issued a Bi-Strategic Commander discussion paper on developing NATO EBAO in July 2007.

This required German feedback. In addition, German commanders in Afghanistan were confronted with the US and the British EBAO approaches. German commanders receive EBAO training seminars, but insiders argued that the missing training for staff officers has caused problems between the Headquarters of the International Security Assistance Force (ISAF) and ISAF Regional Commands. The ISAF Commander has defined his tasks and orders based on the EBAO logic and expects his subordinates to provide feedback within the same framework. In the case of the Bundeswehr, problems arising from the approach of the ISAF Commander were said to be manageable, because the German approach to mission command and Bundeswehr command processes come close to that of EBAO. There are, however, problems between the German and the US understanding of EBAO. Differences result mainly from military culture and the more quantitative US approach, which is seen as problematic by the Bundeswehr. Experience in Afghanistan has thus prompted the Bundeswehr to advance its own conceptual thinking on EBAO.[51] General Mattis's critique of the US approach to Effects-Based Operations,[52] by contrast, did not affect the German position on EBAO. In Germany the Mattis memorandum was interpreted mainly as the expression of dissent between the different US services and not as a critique of NATO's EBAO understanding.[53]

Despite the lack of an overall Bundeswehr concept on EBAO, the German Air Force published a concept paper (*Konzeptionelle Grundvorstellung*) on EBAO in May 2007.[54] With this paper the air force leads conceptual thinking in the Bundeswehr and requires the remaining Bundeswehr services to define their position in relation to the air force.[55] The air force paper describes the system of systems approach that is the basis for EBAO and identifies the role of air power in an effects-based approach. In doing so the paper builds on the six generic capability categories of the Bundeswehr and outlines the contribution of air power to the political, economic, and civil domain of EBAO. In the future, this approach could be used to identify contributions of the Bundeswehr, its services, and other governmental and nongovernmental actors. The concept paper concludes with a discussion of the EBAO-related need for action in the six areas of transformation, thus highlighting what needs to be done in order to make the German Air Force "EBAO ready."

Given its attention to EBAO, Germany actively participates in the Multinational Experiment (MNE) series.[56] The Bundeswehr took part in all MNEs from the beginning and has since become the lead nation for Information Strategy/ Information Operations and for the so-called Knowledge Development (KD).

KD is a method of providing holistic modeling and assessment of information for EBAO in general. The application of the KD process provides planners with a framework for analyzing a situation based on the PMESII domain model,[57] and outlining possible alternative options to achieve a desired end-state. By providing the method for creating EBAO-relevant knowledge, the Bundeswehr plays an influential role in the international EBAO community.

Between June and August 2007 the German KD approach was tested by KFOR Headquarters in Kosovo.[58] The feedback confirmed that the approach helps improve the situational understanding of commanders, in particular with regard to the situation on the spot and interaction between different local parties. Almost parallel to the Bundeswehr's experiment in Kosovo, US armed forces tested their KD approach called Operational Net Assessment (ONA) in Afghanistan during ISAF X. Experimentation results were the basis for Germany's contribution to the NATO Bi-Strategic Commanders' KD Concept,[59] and the draft KD Handbook.

At the domestic front, EBAO must be interpreted in light of the overall goal to advance interagency cooperation through networked security. This is no easy task, as German constitutional law puts strict limits on the domestic use of Bundeswehr assets and thus also on Bundeswehr intelligence. Despite that, the Bundeswehr is thinking of sharing its concepts and instruments developed in the EBAO context with other ministries. There are initial ideas to open the knowledge base for other ministries and to invite other ministries to participate in the Bundeswehr's OHQ in Potsdam. So far, the Ministry of the Interior has sent a liaison officer to Potsdam, and the Ministry of Foreign Affairs provides a diplomatic advisor to the OHQ commander.[60] Similarly, the Bundeswehr, the federal police, and Germany's air traffic control agency, DFS, which is part of the Ministry of Transport, Building and Urban Affairs, use a collaborative information environment to run the national Air Defense Command.

Despite these positive signs, however, there is still a long way to go, since civil ministries seem somewhat reluctant to embrace the Bundeswehr's offers.[61] This explains why Germany, unlike other nations, has not yet seen a comprehensive redesign of its strategic political leadership structure. At least on paper, the Federal Security Council—where the ministers of Foreign Affairs, Defense, the Interior, Justice, Finance, Economics and Technology, and Economic Development and Cooperation meet under the leadership of the chancellor—could serve as the strategic interagency body that could bring networked security to life. In practice, however, the body is not used as it should be. Discussions about how to reinvigorate the council and beef up its institutional capacities have been going on for years.[62] So far no concrete action to remedy the situation is in sight. This is a problem not only for Germany's security policy but also for the future effectiveness of the Bundeswehr. As long as other ministries do not come up with the necessary reforms to make sure that integrated security concepts can be developed and implemented, the political benefits of Bundeswehr transformation will remain limited.

Concept Development and Experimentation

CD&E is the preferred supporting methodology to transform the Bundeswehr.[63] Because today's security risks are complex, there is no one-size-fits-all solution. CD&E aims at identifying promising new conceptual ideas, further developing those ideas toward a certain degree of maturity, and creating defined and reproducible experimental environments in order to optimize concepts in an iterative process of concept development and experimentation. By using modeling, simulation, and other techniques, CD&E can contribute to providing an early assessment of the potential military outcome of new thinking, thereby pointing out intended and unintended consequences. That is why CD&E is the workbench for German defense transformation.[64]

In 2003 the Ministry of Defense adopted a CD&E working program.[65] In addition to the 2006 concept paper on modeling and simulation, a new concept paper on CD&E was adopted in December 2008.[66] This concept paper clarifies CD&E-relevant structures, responsibilities, and processes within the Bundeswehr. To provide CD&E support the Bundeswehr has set up a program of work for experimentation.[67] From 2008 to 2011 this program of work will be expanded to form a fully fledged test and simulation environment. During that time the industry will join as well and fully participate in Bundeswehr CD&E activities.[68]

So far, German CD&E work receives overall guidance from the Armed Forces Staff. Operational work across the services is coordinated by the Bundeswehr Transformation Center. Together with the Response Forces Operations Command, the Bundeswehr Transformation Center and the Armed Forces Staff make up the organizational triangle responsible for Bundeswehr CD&E.[69] Germany's focus on CD&E is also underlined by its strong support for international CD&E activities and its active participation in the Multinational Interoperability Council (MIC) since 1999. Consequently the position of the NATO Deputy Assistant Chief of Staff Joint Experimentation, Exercises, and

Assessment (DACOS JEEA) was held by a German army brigadier from May 2006 to April 2009. The NATO ACT working program on CD&E also provides important conceptual input for German CD&E activities.

Today, the Bundeswehr has an annual CD&E budget of around €16 million.[70] Germany's CD&E activities focus mainly on network-enabled capabilities, advancing interagency interaction, and EBAO. Experiments deal with issues such as the RobCOP, KD, and multinational information operations. C2 processes, instruments, and organizational structures for a distributed operational-level headquarters have been part of the "Home Base" experiment. "Sea-based C2" tests the feasibility of C2 from the sea for a land force contingent. Different aspects of jointness, for example, are evaluated as part of experiments on "Joint Fire Support" and "Urban Operations."[71] In order to advance interagency CD&E activities, the Ministry of Defense coordinates with the federal ministries of Foreign Affairs, Economics and Technology, the Interior, Economic Cooperation and Development, and Transport, Building, and Urban Affairs.[72]

So far CD&E has provided valuable insights into specific areas of work such as the impact of NEC on C2 processes and the interplay between human factors and NEC. CD&E also helped underline the value of new approaches such as those that KD tested in Kosovo. There is, however, room for improvement. First of all, CD&E activities could put more joint focus on NetOpFü and Rob-COP capabilities. Although these issues are addressed in CD&E experiments, many experiments are built bottom-up from an individual service perspective rather than top-down from a Bundeswehr joint perspective. Interviewees also underlined that the link between CD&E, training and education, and procurement could be strengthened. In the latter case, processes could be accelerated by fully exploiting the potential of CD&E during the initial stages of procurement planning.[73] This would be an important step to increase procurement flexibility. Finally, more could be done to implement CD&E results via UOR in order to bring added value to the front line more quickly. Insiders argue that this idea is currently being advanced by Bundeswehr planners. In addition, it is also said that there will be a stronger CD&E focus on achieving initial NetOpFü capabilities in 2012/13.[74]

TRANSFORMATION SCORECARD

The Bundeswehr transformation scorecard is mixed. On the one hand it can be said that the basic idea of transformation has been established and has started to influence all services of the Bundeswehr. This has also helped the Bundeswehr to assume leading roles in some areas that are relevant for transformation. In addition, new mission and transformation requirements are also strongly influencing training and exercises. On the other hand, there are deficits and therefore substantial room for improvement with regard to leadership, human resources management, and communication, as well as funding and procurement.

Achievements

Each of the three main Bundeswehr services has started to launch substantial transformation-relevant initiatives. The air force has championed the cause of EBAO in the Bundeswehr with an innovative concept paper. The air force is also the driving force behind exploiting space-based capabilities to advance command and control as well as intelligence, surveillance, and reconnaissance.[75] In line with this, Germany has also opened the Joint Air Power Competence Center[76] as a NATO Center of Excellence and the national interagency Air Defense Command. The navy is preparing to become expeditionary with a focus, inter alia, on protecting sea lines of communication and providing sea-based support for military operations on shore.[77] In light of these changes the navy has adopted a new concept paper on using the sea as a basis for the conduct of joint operations,[78] and has opened a NATO Center of Excellence for Confined and Shallow Waters.[79] The army has defined several beacon projects such as Joint Fires, the fielding of new reconnaissance units, and an air-mobile brigade as well as action to improve C2 interoperability for multinational operations.[80] The army has established a NATO Center of Excellence for Military Engineering.[81] Furthermore, the RobCOP is the key reference project to advance Bundeswehr jointness. Thanks to the overall concept of networked security, some ideas have also traveled to other actors of the German security sector, thereby contributing to improving interagency interaction in different areas such as intelligence exchange (for example, the Joint Counter-Terrorism Center between the Federal Criminal Police Office and the Federal Office for the Protection of the Constitution), homeland security (for example, the Joint Maritime Center with the participation of federal and Länder agencies), or stabilization operations (for example, Provincial Reconstruction Team).

In the field of training and exercises there is considerable attention to the new demands of ongoing international operations and NetOpFü. In order to improve mission preparedness, joint training programs and courses as well as new tactics, techniques, and procedures receive priority treatment.[82] In addition, the Bundeswehr also puts more emphasis on issues such as intercultural competences, communication, foreign language expertise, and information about foreign political systems. There is also a strong emphasis on psychological aspects, such as "culture clashes" resulting from operations in mission environments that were hitherto unknown to German soldiers. Normative education in the sense of what values the Bundeswehr is fighting for also becomes more important.[83] Besides testing NetOpFü-relevant applications in exercises, the Bundeswehr services also look at the impact of NetOpFü on training needs (for example, implications on knowledge management, leadership support, and collaborative teamwork).[84] Furthermore, the Bundeswehr interprets modeling and simulation as well as e-learning as two important instruments to advance training and education. Therefore the Bundeswehr's main training centers will be included in the new IT Directorate's test and simulation environment, which will be established in the future. To that purpose the capabilities of existing battlefield simulation centers of the Bundeswehr will be beefed up.[85] In addition, the use of e-learning will be expanded in order to provide members of the Bundeswehr with tailored education programs wherever they work.[86]

Shortfalls and Room for Improvement

The first noticeable problem is the lack of powerful transformation promoters. As will be argued in the last section of this paper, this is more a problem of leadership and will, rather than of organizational reform. From an organizational perspective, processes and structures have been reorganized to address the challenges of transformation in light of a networked security approach. But ardent zeal to use these processes and structures in order to implement changes with demanding and transformation-focused requirements seems lacking.

At the beginning of 2003/4 German defense transformation had visible promoters. The Chief of Defense together with the newly established Bundeswehr Transformation Center and the Armed Forces Staff were pushing for transformation, but today, the Bundeswehr transformation process lacks further promoters. This has prompted many to believe that Bundeswehr transformation has run out of steam and come to a standstill. It is unclear what caused the loss of leadership. It may be speculated that setbacks in NATO, the change of leadership in the Pentagon, and internal difficulties in getting transformation to the fore under substantial financial limits have cushioned the Bundeswehr transformation process. Overall, the detrimental effects of the loss of promoters are reinforced by the lack of guidance for implementation that was discussed in detail above.

Second, transformation puts a premium on the constant need to adapt to new circumstances. This asks for systematic change management that starts with a vision, aligns people, organization, and technology, and is based on effective communication. While the vision has been established, shortfalls in the remaining two areas are very problematic. In particular, change management requires people to be much more flexible and to have the necessary capabilities to cope with new situations. Successful change management depends on effective human resources management. But in this area the Bundeswehr still has a lot of work to do. Joint human resources management has only just started with the establishment of joint personnel management for noncommissioned officers. The services still "own" their people. They are in charge of human resources management and management development despite the fact that around 50 percent of the staff officers work in joint positions inside the Joint Support Service. Personnel administration puts additional ties on people and is detrimental to advancing flexibility.[87] Finally, there is the long-term trend of demographic change that challenges the basis of conscription. This challenge has been recognized but not yet adequately tackled. In the field of human resources management the Bundeswehr could learn from other countries. The Netherlands, for example, is an interesting case. There, changes from conscription to professionalization and growing international engagements have led to a new human resources approach that builds on a single human resources policy and a joint human resources pool that is managed by the human resources department.

A third problem was, and to some extent still is, poor strategic communication. The German Ministry of Defense never really managed to develop a strategic rationale to communicate why and how the Bundeswehr needs to transform. Unfortunately, one of the first concrete manifestations of change was the new basing concept adopted under Defense Minister Struck that led to base closure and uncertainty among Bundeswehr employees. Since then, transformation has had a negative connotation. The same also holds true for communication with the German Parliament. In this regard, the Ministry of Defense and the German government did not succeed in establishing a clear case for the Bundeswehr that might have helped find new ground for explaining why the Bundeswehr is active abroad, what the armed forces are doing there, and why all this is important for Germany's national security and Germany's position in international politics.

Finally, there is a funding problem that leads to procurement shortfalls.

Poor communication and the lack of political will can explain why slashing Bundeswehr funds has become a key means to restructure the federal budget since the end of the Cold War. Negative consequences of inadequate funding are reinforced by the fact that large parts of the investment budget of the Bundeswehr are eaten up by big platforms (for example, Eurofighter, A400M, and F125) that will enter the services in the years to come. The need to finance these platforms leads to drastic cuts in smaller programs that are sometimes more important to give the Bundeswehr the much needed flexibility to react to unforeseen events.[88] So far, the Ministry of Defense expects the defense industry to meet existing contractual obligations even if procurement projects are massively delayed (for example, A400M). Other countries, by contrast, seem more prepared to cut into big platform projects in order to expand conceptual flexibility and free up resources.[89]

Right now the Bundeswehr's conceptual ambition and available funds are out of balance. The need to meet operational requirements further widens this gap, which threatens the credibility of defense transformation.[90] This also endangers the Bundeswehr's preparedness for international operations. In this respect the Bundeswehr Plan 2008 leaves no doubt that there are significant limitations for international operations in the higher, more demanding intensity spectrum. Therefore, missions in the full spectrum of current operational requirements will not be possible until long after 2012.[91] Financial pressure, however, could also serve as a stimulus for long-needed changes. In times of networked security, international operations should be interpreted as an all-government task requiring an all-government approach to funding. At the moment, international operations are covered by the defense budget, but ideas have been circulated to finance international operations out of the general budget of the federal government. In addition, the use of national and international capital markets could be an alternative route to Bundeswehr financing.

EXPLAINING BUNDESWEHR TRANSFORMATION

Three of the four strands of literature discussed in the chapter by Farrell and Terriff are important to explain the process of military transformation in Germany. The discussion about strategic culture serves as the general background against which the development of German security policy since the end of the Cold War must be interpreted. Most important are reflections about military innovation, which depend to a very large degree on the notion of strategic culture. Norm diffusion is also relevant, but in a somewhat different way than suggested by Farrell and Terriff. In the case of Bundeswehr transformation, alliance theory is less relevant and will thus not be discussed.[92]

Strategic Culture

Among the different "unquestioned orientations towards and the assumptions about the political world" that denote political culture,[93] Germany's understanding of international challenges and its responses since the end of World War II were influenced by at least three ideas. First, Germany learned to regain sovereignty as a nation by transferring powers to the European Union. This process went hand in hand with a very strong belief in the benefits of multilateralism and the need to support multilateral institutions. This explains why the Bundeswehr is tightly integrated into international military structures and multinational units. It also illustrates why "going national" in the foreign and security realm has never been a real option for Germany.

As a direct consequence of the positive experience with multilateralism, international solidarity is a key driver for German activities in the fields of foreign, security, and development policy. The "international cause" helps justify national action. This explains, for example, German participation in the EU mission in the Congo to strengthen Europe's security and defense policy, and Berlin's request, in cooperation with The Netherlands and Canada, to have NATO take over the leadership of ISAF. This, however, comes with a price. So far Germany has hardly felt a need to explain and justify its national interests in foreign and security terms. Given the country's position as one of the world's leading export nations, that is surprising. As a consequence, there is no grand strategy for German foreign and security policy. In the context of the networked security paradigm this could turn out to be a problem because there is no guidance to align the responsibilities and capabilities of the different ministries with overall national interests.

Finally, as a post-heroic society,[94] Germany has a natural inclination toward diplomatic and economic, rather than military, power. Since 1990 and, more important, September 2001, political parties in Germany have failed to explain what these changes imply for Germany's future role in the world and for the instruments of power it needs to stand up for its interests.[95] Although German public opinion favors a more active international role, there has been no public discussion about the ways and means to play that role.[96] There is thus no "strategic rationale" on which the Ministry of Defense or the Bundeswehr could ground their actions. As a result, the Ministry of Defense and the Bundeswehr are most often agenda-takers, not agenda-setters. And if the Ministry of Defense or the Bundeswehr takes the lead on an issue such as networked security, for example, it is most likely to meet with reservation or indifference.

Military Innovation

Looking at organizational culture is a very fruitful approach to explain the outcome of Germany defense transformation. Right from the beginning, the Bundeswehr has tried to replace the technocentric US approach to NEC with a more comprehensive understanding that balances technology, people, and organization. The Bundeswehr's belief that military innovation is more than technological progress can be dated back to Prussian reformers of the eighteenth and nineteenth centuries. Already at that time, military innovation was seen as part of a more comprehensive need for societal reform to meet the challenges of those days. Many of the leading reformers at the time were military thinkers who took part in societal and political debate about the future of Prussia.[97] Those reformers have influenced German military strategic thinking until this day. It is thus not surprising that the Bundeswehr opted for a comprehensive NEC approach and the need to embed military transformation in the broader framework of networked security.

Another historical legacy is more problematic. With the end of World War II, German military tradition in the positive sense was broken. Since then there has been a general suspicion vis-à-vis armed forces.[98] Germany maintains armed forces with the goal of keeping them away from genuine military tasks, one commentator recently wrote.[99] This observation is correct in the sense that Germany has not yet developed a risk-taking security culture like the former European colonial powers. Today the Bundeswehr operates under close public and political scrutiny. This means that the security of Bundeswehr soldiers abroad is at least as important as their readiness to accomplish a given mission. The duty to avoid everything that could risk the lives of German soldiers, German political scientist Herfried Münkler observed, has become an implicit permanent task for the Bundeswehr. Needless to say, that attitude is limiting the Bundeswehr's effectiveness.[100]

The German problem with leadership is another aspect that needs to be taken into account. Again as a result of history, there is no real center of power in Germany. The chancellor, for example, has the power to provide guidance to her or his ministers (*Richtlinienkompetenz*). Germany, however, has a corporatist and consensus-oriented political culture. The government is a team that works together. The chancellor, despite her or his formal authority, is only first among equals. Therefore German ministers enjoy a very large degree of autonomy to run their ministries, which is guaranteed by German constitutional law. These characteristics of the political system are reflected in the military leadership of the Bundeswehr. The position of the Chief of Defense, Bundeswehr is comparable to that of the Chancellor in the sense that he is also first among equals. His powers vis-à-vis the Services Chiefs were strengthened with the 2005 Berlin Decree, but he is reluctant to use this authority, because he favors consensus over rivalry.

Finally, the role of the German defense industry is somewhat ambivalent. Contrary to the UK, The Netherlands, or Sweden, Germany like France has so far preferred national defense industrial consolidation over international mergers and acquisitions. This becomes more and more problematic. As a consequence of shrinking defense budgets, the German home market loses strategic relevance for German defense companies. These companies depend ever more on exports, which means that it takes extra effort to convince these companies to maintain defense industrial capabilities that are of importance for the Bundeswehr. In addition, a very large part of the German defense investment budget is committed to the European Aeronautic Defense and Space Company (EADS). The current problems with EADS thus also limit the leeway of the Ministry of Defense and the Bundeswehr in particular with regard to an eventual radical shake-up of procurement projects that could help free up scarce resources. Unlike other countries, in Germany the defense industry is thus less of a driving force for defense transformation. There is, however, a sign of hope with the creation of the Open Community, an association of defense companies that wants to promote open architectures and nonproprietary standards for defense solutions.[101] Admittedly, the influence of these companies on the Bundeswehr ultimately depends on the number of projects they receive in the future.

Norm Diffusion

Changing security challenges are a key motive for German defense transformation. New security challenges require not only different armed forces but also new approaches to mix military and nonmilitary capabilities. As has been argued, EBAO resonates well with the German strategic culture. The need to change the Bundeswehr, by contrast, was accepted only reluctantly. The message "Either you transform, or you are no longer relevant" has been supported by some exponents of the Bundeswehr, but it seems that German politicians have never really bought this idea. In the US, where key concepts were developed, transformation was meant to improve armed forces for high-intensity war-fighting. This idea was resisted in Germany. Therefore, key transformation concepts faced acceptance problems in Germany. The lesson to be learned from this is that role models for force transformation have clear limitations. International organizations that can assist countries in transforming their armed forces are thus well advised to cover the whole spectrum of military tasks and to underline that the principles of transformation are as relevant for stability operations as they are for intervention operations.

This argument goes hand in hand with the distinction between process and content norms, advanced in the theoretical chapter by Farrell and Terriff. If one interprets the process norm as describing the general institutional setting in which the ideas of the content norm come to play, then this author believes that the process norm is more important for Bundeswehr transformation than the content norm. In the US, the United Kingdom, France, or The Netherlands, defense transformation takes place against the background of a clear political vision about the use of force and the tasks that can be accomplished with military power. This leads to a—more or less—coherent vision of the armed forces and guides force transformation. In Germany there is no such link. Still today, the political establishment has a hard time defining German interest and explaining in public why armed forces are maintained. In this context, debates on structural change (process norm) seem more digestible. But this case study also showed that process and content norms are intertwined. This is reflected in the reluctance to implement structural changes in Germany. Again, the experience of other countries is telling. In The Netherlands, the political readiness to use armed forces as a fully fledged instrument of power has led not only to the abolition of conscription but also to sweeping structural changes at the top of the armed forces and the Ministry of Defense. Germany has followed a similar path, but much more cautiously and less forcefully. Whether it will go "the last mile," for instance by redistributing powers between the Chief of Defense, Bundeswehr, and the Service Chiefs, very much depends on the will of the political and military leadership.

6 Innovating on a Shrinking Playing Field: Military Change in The Netherlands Armed Forces

Rob de Wijk and Frans Osinga

This chapter traces the trajectory of changes in the Dutch military of the past two decades. It aims to disaggregate the complex set of drivers of change and assess which factors have been dominant in that process. On the face of it, in the outward appearance of the Dutch armed forces and in the declared defense policy, the influence of NATO and its Transformation initiative seem easily identifiable. Already in the 1990s, the Dutch armed forces were among the first European militaries to embark on a radical restructuring process that, in hindsight, can be regarded as transformation in NATO parlance. As the most recent 2007 White Paper proudly asserts: "[In] less than 15 years it has evolved from a large conscript based force into a force that ranks among the most modern in the world, with professional, highly motivated personnel operating state of the art equipment, ready to promote peace and security world-wide. It is capable to play an active role in all phases of a conflict, including those at the highest end on the spectrum of violence, but always alongside our allies," it notes, reflecting the importance of NATO.[1]

Two Dutch analysts underwrite NATO's influence, stating that "the shape and structure of the Dutch armed forces have been strongly affected by NATO due to the long and loyal membership to the alliance." They attributed this in large part to the NATO Defense Planning Process and to the extensive and deep operational integration of Dutch armed services into NATO command structure and force structures.[2] Indeed, for conducting effective operations, there is a substantial measure of dependency on cooperation with, and contributions by, NATO partners, they add.[3] The White Paper too seems to confirm the centrality of NATO by discussing the ambitions of NATO, the importance of burden sharing, and reinforcing European military capabilities, the commitment to the NATO Response Force (NRF), as well as the aims of NATO transformation. This includes specific mention of the outline of NATO's 2007 Defense Requirements Review. Also reinforcing the link with NATO Military Transformation is the explicit discussion on the relevance of Network-Enabled Capability (NEC), interoperability, and the requirement for additional investments in those capabilities. The need for continuous modernization and accelerated lessons learned processes is addressed in the statement that acquisition procedures must be changed to accommodate emerging and urgent operational requirements, so-called fast track requirement and "fast track procurement" procedures.

Moreover, the expeditionary nature of Dutch armed forces is demonstrated in the numerous operations it has participated in, in areas such as Iraq, Bosnia, Kosovo, Sudan, Congo, Liberia, the Arabian Sea, the Persian Gulf, the Mediterranean, and Indonesia. It has deployed a sizable contingent to Afghanistan, involving stability and reconstruction missions in addition to very intense counterinsurgency operations. It deploys 6,000 military personnel on a yearly basis, not counting commitments to the NRF and EU Battle Groups. It emphasizes high-quality, high-tech weapons systems that are relevant in a variety of conflicts. Operating on a limited annual budget of about €8 billion, it maintains a deployable force of around 30,000 personnel, C-130 Hercules and KDC-10 transport and air-to-air refueling (AAR) aircraft, a sizable fleet of transport helicopters, 24 Apache combat helicopters, Patriot SAM systems equipped with Tactical Ballistic Missile (TBM) defense capabilities, 8 frigates, 4 submarines, 2 amphibious transport ships, 3 brigades, and 88 Leopard II Main Battle Tanks (MBT). In the near future a decision is expected to acquire Joint Strike Fighters (JSFs) to replace the current fleet of 108 F-16s. Indeed, the Dutch armed forces display many features one would expect of a "transformed" military, and the current Netherlands defense policy seems quite in accordance with the NATO transformation narrative and agenda.

However, the same 2007 White Paper also offers insight into the deeper underlying dynamics driving the changes within the Dutch armed forces. Reflecting on the evolution of the Dutch armed forces during the past two decades, and describing the expectations for the coming years, in the final paragraph it notes that "this process was marked by a continuous effort to maintain a balance between ambitions and available means. Drastic measures to achieve that balance have not been avoided. Our priorities for the coming years lie with personnel and the improvement of the operational 'deployability' and usability of the armed forces. Also in light of ongoing operations this will require additional financial re-shuffling within the existing budget. In addition, in order to maintain a balanced budget, several plans have been canceled, others will be postponed, and some weapon equipment will need to be sold." This suggests that in addition to NATO, other factors—money and domestic politics, for instance—have been very influential in shaping Dutch defense policy and force structure.

This chapter aims to unravel the mix of shaping factors in Dutch defense policy and of the Dutch armed forces. A brief introduction of some dominant and enduring tenets of Dutch foreign and defense policy will serve to illustrate the cultural *couleur local* of the debates on defense policy. A chronological overview of the major policy-driven developments since 1990 will highlight the changes in missions and force structure of the armed forces toward an expeditionary orientation. Sections three and four examine the diverse trajectories of, and the internal change processes within, the different armed services, and address the question to what extent ideas such as Effects Based Operations (EBO)—or its derivatives, the Effects Based Approach to Operations (EBAO) and the Comprehensive Approach—and Network Centric Warfare (NCW) took hold. The final section will draw conclusions concerning the processes that marked military change in the past two decades. In brief, it will paint a picture of a continually shrinking playing field where civilian intervention dictated that the main game would be expeditionary operations, and the services were left with deciding what sort of field players could be afforded and what tactics should be used.

A SMALL COUNTRY'S STRATEGIC CULTURE

Force transformation can only be understood in the context of the tradition of The Netherlands foreign policy—that is, its strategic culture.[4] Very frequently those traditions manifest themselves explicitly in defense white papers and speeches by ministers of defense, in which they are presented as core foundations and guiding principles underpinning security and defense policy.[5] The Netherlands is a medium-size industrial power ranking tenth on the OECD list of wealthy countries, and its foreign policy shares characteristics with that of similarly highly developed, rich, medium industrial powers, such as Canada and the Scandinavian countries. Dutch foreign policy was once characterized as one based on "peace, profits and principles,"[6] which indeed marks some recurring and interlinked approaches or traditions in Dutch foreign policy. The first is a strong legal approach. This goes back to Hugo Grotius, who in the seventeenth century was one of the founders of international law. Dutch interest in international law has remained constant over the centuries. As a trade nation, The Netherlands always attached great value to a strong international legal order as an instrument to create stability. This results in a status quo orientation. Its size and commerce-based economy also inspires its attempts to keep its independence from the major continental powers by promoting coalition formation so as to balance against specific great powers. This underpinned the support for the Pax Britannica and the postwar support for the Pax Americana and the emphasis on good transatlantic relations.

Second, and following from this, The Netherlands has become a strong proponent of international organizations, in particular since the end of World War II. Institutional and legal approaches were and are considered complementary. The Netherlands was among the founding members of the West European Union, the forerunner of the present day European Union, NATO, the UN, the OSCE, the World Bank, and others. As a political and military principle, no military task whatsoever will be carried out unilaterally. Dutch soldiers will always be employed in the context of NATO, the UN, or EU, and after sanctioning by those bodies. To this day, transatlantic relations and consequently NATO are seen as the cornerstone of Dutch foreign policy. Even its support for European Security and Defense Policy (ESDP) is in part inspired by the fact that improving Europe's military capabilities will address also the issue of burden-sharing within NATO.

The third tenet follows from the previous two: a certain aversion to military matters and the use of force. Dutch society has come to regard war, like slavery, as "subconsciously unthinkable," as John Mueller has termed the "de-bellicozation of the Western world" (and indeed he calls it the "Hollandization" phenomenon to note the origins of where this attitude first took hold).[7] Instead there is an emphasis on international law, and multinational organizations are the preferred instruments to create an orderly world. In a society marked by a political culture of consultation, negotiation, accommodation, consensus, and compromise, engaging in war and using military force to resolve a conflict does not come naturally. Indeed, during the 1960s and 1970s pacifism had a relatively large influence on Dutch society.[8] While it has lost most of its symbolic impact in the post–Cold War days, it goes some way in explaining that, in a broad sense, security is now considered in political, social, economic, ecological, and humanitarian terms. Risks and conflicts are always seen as multifaceted, having many interdependent causes, and security can be obtained only by the interplay of several instruments, military and nonmilitary (economic, social, educational, diplomatic, development, and so forth). Moreover, conflicts and tensions in remote regions can have spill-over effects and with modern-day interdependencies will result in security risks at home.[9] Finally, a sense of responsibility to protect humanitarian principles inspires defense and security policy, rather than the prospect of waging war for territorial defense or the protection of naked national interests.

In those features and developments the strategic culture in The Netherlands matches the prevailing European strategic culture, which has been described as a set of norms that bound strategic thinking: coercion and not brute force is necessary; force must be used legitimately—that is, in a multilateral manner; it must have motivation that is seen as just; special attention should be paid to force protection and limiting collateral damage; finally, wars must be short.[10] In keeping with these developments toward a "postmodern" or "postmilitary" society,[11] one can describe the Dutch military as it has developed over the past three decades as a "postmodern" one.[12]

A final tenet—or tension rather—that followed from this political climate has been the unresolved question as to what extent The Netherlands armed forces should retain the capability to conduct high-intensity combat operations, followed by the question of whether a focus on peacekeeping operations would be more appropriate for a country the size of The Netherlands. These two positions nicely overlap the left and right end of the Dutch political spectrum: the leftist parties promoted the idea to focus the armed forces on humanitarian missions, de-emphasize the war-fighting orientation, and enjoy the resulting "peace-dividend" (to be used for health care, social security, and education); the right (and the three services) backed the idea of having a force that could conduct more demanding and robust combat operations that could also de-escalate and conduct peace operations.[13] The series of policy papers of the 1990s discussed below reflects this enduring tension.

TOWARD AN EXPEDITIONARY POSTURE
Downsizing

In The Netherlands, defense policies are derived from the country's foreign and security policies, which are the responsibility of the minister of foreign affairs.[14] By tradition, the prime minister is the chairman of the council of ministers and has little power in comparison with his British counterpart. Moreover, there is no national security doctrine guiding foreign and security policy. Major changes in the Dutch force posture are announced in white papers and letters to Parliament, minor changes in letters to Parliament as well, and in the memorandum of the annual defense budget. The minister of foreign affairs is responsible for the political part of a white paper, providing the necessary guidance for the further development of the tasks, roles, and missions of the armed forces. Modernization and transformation are part of the defense planning process that is the responsibly of the Chief of the Defence Staff, who receives political guidance from the minister of defense.

Current developments and trends within the Dutch armed forces must be seen as part of a trajectory that commenced during the first half of the 1990s (see Table 6.1 for a brief summary of major policy decisions and budget trends from 1990 to 2007). In that period the Concepts Division of the Defence Staff, together with the Political Affairs Department, took the lead in the restructuring process. The process started soon after the collapse of the Berlin Wall with a "strategic vision" document.[15] During the 1990s the force transformation process was largely an internal affair. Importantly for this study, this process took place despite objections by NATO. Although NATO was working on a new Alliance *Strategic Concept,* the Alliance's leadership was afraid to draw quick and far-reaching conclusions from the new security situation. During the first half of the 1990s, NATO actively tried to slow down the force transformation process. Subsequently, senior military and civilian officers traveled up and down to Brussels to explain The Hague's position.

The Netherlands's position as a front-runner can be explained by the fact that Parliament traditionally focused primarily on the budget and procurement, and was less interested in conceptual matters regarding the armed forces. As a matter of fact, Parliament lacked military knowledge and accepted major changes in the defense posture without too much debate or effort to influence the trajectory. Parliamentary interest concerned primarily the creation of a "force for good." The character of this force was the topic of parliamentary debate, a debate that has up to today not subsided. The left-wing parties argued that The Netherlands should be reoriented toward a peacekeeping force, whereas the right side of the political spectrum supported the idea of maintaining a force that could also conduct more robust combat operations. This relative lack of detailed parliamentary intervention resulted in substantial latitude for the MOD in regard to the force structure. MOD defense planners supported the idea of a war-fighting force capable of executing peace support missions.

TABLE 6.1

Overview of Major Policy Decisions and Budget Trends

Year	Source	Cause	Changes in Security Situation	Most Important Measures for Armed Forces	Annual Defense Budget in Billions of Euros		Actual Budget in % GDP[c]	NATO Avg. Budget in % GDP[d]
					Estimated[a]	Actual[b]		
1990					6472.6	6429.7	2.64	Avg. period 1990–94: 3.5
1991	Defence White Paper (Defensie Nota 1991)[e]	Dissolution of the Warsaw Pact as a military organization; withdrawal of Soviet troops from Eastern Europe, cost reductions	Lengthened readiness requirements in case of a full-scale attack	KM: Reduction of personnel 15% up to 1995 and 10% thereafter, divestment of 2 T-class submarines, 4 mine hunters, and 2 S-frigates KL: Increased flexibility and mobility of divisions,/ divestment of 40% of the armed vehicles and artillery, KLU: Reduction of fighter jets from 162 to 144, Reduction of personnel 16%, closure of Ypenburg Air Base	6446.4	6431.5	2.50	
1992					6392.4	6613.0	2.46	
1993	Priorities Paper 1993 (Prioriteitennota 1993)[f]	Dissolution of the Warsaw Pact and the Soviet Union; pro-Western position of Eastern European countries; crisis management outside the NATO area, cost reductions	None	Decision to suspend conscription so that as of 1-1-1998 all armed forces are voluntary KM: The disbandment of a Task Force and divestment of its 4 frigates, total personnel to be restructured to 17,500 KL: Personnel restructured to 36,000 KLU: Personnel restructured to 13,000	6395.0	6290.8	2.28	
1994	November letter (Novemberbrief 1994)[k]	Financial Constraints	None	Efficiency measures and emphasis on international cooperation	6110.4	6250.1	2.15	

Year	Policy document	Threat/Event		Measures				Average period
1995					6133.7	6089.2	1.99	1995–99: 2.7
1996	Annual Defense Budget 1996–1997[i]		None		6169.7	6220.3	1.95	
1997				KM: Increased reduction of personnel, total of −20% since 1990, delivery of 8 M-frigates KL: Reduction of personnel of 52% since 1990. Improvement of existing material, e.g. tanks and short-range air defense KLU: General restructuring, e.g. through international cooperation	6191.5	6331.4	1.85	
1998					6338.9	6347.7	1.75	
1999					6416.6	6786.4	1.76	
2000	Annual Defense Budget[j] Defence White Paper 2000 (Defensienota 2000)	Restricted NATO defense tasks, cost reductions	None	KM: Sale of mine hunters and Orions and purchase a second amphibious assault ship KL: Increase of mechanized infantry, divestment of 90 Leopard II tanks KLU: Staff increase, divestment of 29 F-16 fighter jets	6439.4	6729.9	1.61	2.6
2001	Action Plan[k] (Actie Plan Terrorismebestrijding en Veiligheid)	Assaults of September 11th	Catastrophic terrorism	Installment of task force to investigate the role of the Defense Department in combating terrorism, analysis and expansion of BBE capacity[l]	6616.7	7192.2	1.61	2.6
2002	Autumn Letter 2002[m] (Najaars-brief 2002)			Reduction of 4,800 jobs, Dollar rate / ISF related benefits, 30 mln. KM: Divestment of 2 L-frigates	7017.3	7358.8	1.58	2.7

TABLE 6.1—cont.

Year	Source	Changes in Security Situation	Cause	Most Important Measures for Armed Forces	Annual Defense Budget in Billions of Euros		Actual Budget in % GDP[c]	NATO Avg. Budget in % GDP[d]
					Estimated[a]	Actual[b]		
2003	Restructuring Plan (Prinsjesdagbrief 2003)[n]	None	Financial constraints, NATO's Quality over Quantity requirements	Reduction of additional 3,800 jobs KM: Divestment of 10 (all) Orions, 2 M-frigates, 2 mine hunters	7311.7	7403.9	1.55	2.7
	Annual Defense Budget 2002–3[o]			KL: Additional 30 mln. in 2004 up to 100 mln. and after that structural increase for participation in peace operations, 3 additional companies of armed infantry, additional 39 howitzer 2000s, divestment of 90 Leopard tanks, 6 Apaches, and the MLRS KLU: Divestment of 29 F-16 fighter jets, all Hawks, dismantling of barracks in Ede and Airbase Twenthe				
2004					7667.9	7607.8	1.55	—[p]
2005	Naval Study 2005[q]	None	Consequences of the Restructuring Plan for the Naval Forces.	KM: Rendering the naval forces more suitable for supporting long operations, divestment of 4 M-frigates, increased mine hunting capabilities, purchase of ocean patrol vessels	7673.2	7742.1	1.52	
2006	Update Restructuring Plan 2003[r] (Actualisering Prinsjesdagbrief 2003)	None	Considering consequences of reorganizations	Improving interoperability with Allies through NEC KL: Appoint a Counter-IED Task Force, Clustering operational intelligence services to ISTAR[s] KLU: Divestment of 6 Apache helicopters	7768.7	7940.1	1.49	

2008 (Beleidsbrief 2008 Wereldwijd Dienstbaar)	means for crisis management operations.	personnel, increased intelligence KL: Divestment of 28 Leopard II tanks, 12 howitzers KLU: Divestment of 18 F-16 jet fighters		
2008	8094.7	–	–	

[a] The numbers for the estimated defense budget (200x) are taken from the Budget of State "Rijksbegroting" of that year (200x), Table 2.1 "De uitgaven en niet-belastingontvangsten."

[b] The numbers of the actual budget of 200xx derive from the Budget of State two years later (of 200x + 2) with the exception of any budget after 2004; those are taken from the Budget of State of the following year (200x + 1).

[c] The Dutch GDP on an annual basis is retrieved through http://www.cbs.nl; the %GDP of 2005 and 2006 are based on estimations, the GDP of 2007 is not known.

[d] See http://www.nato.int/issues/defence_expenditures/index.html.

[e] TK 1990–91, 21991, nr 2–3.

[f] KLU: Koninklijke Luchtmacht (Royal Netherlands Air Force); KM: Koninklijke Marine (Royal Netherlands Navy); KL: Koninklijke Landmacht (Royal Netherlands Army).

[g] TK 1992–93, 22975, nr. 2.

[h] TK 1993–94, 23 900 X, nr. 8.

[i] TK 1996–97, 25000 X, nr 1.

[j] TK 1999–2000, 26800 X, nr 2.

[k] Letter of 5 October 2001, ref: 512537/501/RD.

[l] BBE: Special Service Unit, Bizondere Bijstandseenheid, a division of the Dutch Navy that specializes in antiterrorism that in 2006 combined with similar forces from the police into the DSI, Service Special Interventions.

[m] Letter of 8 November 2002.

[n] Letter of 2 June 2006, ref: HDAB20061808 5.

[o] TK 2002–3, 28600 X, nr 2.

[p] Because of disagreement about NATO's new defense expenditure definitions, a NATO average since 2004 cannot be given. See http://www.nato.int/docu/pr/2005/p050609e.htm .

[q] De Marinestudie 2005, 14 October 2005. See http://www.mindef.nl/binaries/marinestudie_tcm15-46345.pdf .

[r] Letter of 2 June 2006, ref: HDAB20060180 85.

[s] Intelligence Surveillance, Target Acquisition and Reconnaissance.

[t] Letter of 18 September 2007. See http://www.mindef.nl/actueel/parlement/kamerbrieven/2007/09/ 20070918_beleidsbrief.aspx.

Over the course of half a decade an expeditionary war-fighting force was developed that was politically sold as a force for good.

Another explanation is that, by coincidence, reformists led the Ministry of Defense: Social-Democratic minister Relus ter Beek and the Chief of the Defence Staff, Arie van der Vlis, and his successor, Henk van den Breemen. They empowered the Concept Division to develop innovative ideas that were generally supported by the political directorate of General Policy Affairs. Soon after the first signs of a collapse of the Warsaw Pact, defense planners in The Hague concluded that there was no longer a need for large active and reserve forces. This warranted force reductions, both in active and war-time strength. That was welcomed by politicians who were interested in force reductions rather than transformation. Following the collapse of the Warsaw Pact, they eagerly collected the "peace dividend." As a first step the government decided to cut back the budget by 2 percent. This was the start of continual budget cuts.

The 1991 Defense White Paper, the *Defensie Nota*, was the first of a series of key political documents guiding the transformation of The Netherlands's armed forces.[16] The white paper emphasized mainly downsizing. The defense of the NATO area was the Cold War mission of the armed forces. The Royal Netherlands Army (RNLA) and the Royal Netherlands Air Force (RNLAF) contributed to the defense of the North German plain, while the Royal Netherlands Navy (RNLN) defended transatlantic Sea Lines of Communication (SLOCs). The structure of the armed forces was geared to these tasks. The army comprised of an army corps with three, largely mobilizable divisions. The core of the land forces were tank battalions, armored infantry brigades, and artillery. Fully integrated in, and assigned to, the NATO command structure, the air force's mission was air defense and support of land forces, the navy's surface ships focused on anti-air warfare and, with submarines, on anti-submarine warfare, in particular in the Atlantic Ocean.

Developed while the Soviet Union still existed, the vision developed by the Concepts Division took as a starting point the assumption that a large-scale surprise attack on NATO territory was still possible, but only after a long preparation time. This warranted a shift from active to mobilization force structure and modest downsizing of the overall strength, which also dovetailed with the 1990 agreement on Conventional Forces in Europe. The white paper considered flexibility, mobility, interoperability, and multifunctionality to be the guiding principles for force restructuring. A modest decrease in the size of the armed force by 16 percent was announced. If the security situation continued to develop favorably, a further 10 to 18 percent cutback was envisioned. This amounted to the disestablishment of one of the three army divisions.

That dovetailed nicely with the Alliance's decision in December 1991 to restructure NATO military capabilities into Reaction Forces and Main Defense Forces.[17] Because of the reduced size of NATO's armed forces, the layered defense concept was no longer sustainable. SACEUR's Counter Concentration Concept required member states to increase mobility to quickly deploy land forces to break through areas at the border with the former Warsaw Pact. Flexibility was emphasized to fight and win on a thinned battlefield. The white paper announced the creation of a rapidly deployable air mobile brigade to fulfill SACEUR's requirement. This included the acquisition of Apache combat helicopters, and Cougar and Chinook transport helicopters that would be operated by the air force. The air force would also procure extra transport aircraft, including air-to-air refueling capacity for carrying out missions outside NATO's Central Region, but at the cost of deactivating about 30 percent of its F-16 fighter force, reducing it to 166.

Capabilities-Based Planning

The 1993 White Paper, or *Priorities Paper*, followed the demise of the Soviet Union.[18] Until then, the force posture of The Netherlands was focused on the defense of NATO territory. Again the process started with a vision document written by the Defense Concepts Division of the Defence Staff. The new document concluded that although a strategic attack on NATO territory was unlikely, new risks were emerging. The Balkans and the Gulf regions had become permanent sources of unrest. The danger of proliferation of weapons of mass destruction and their means of delivery was real, and international terrorism and international crime were considered a growing threat. These risks, together with the scarcity of natural resources, threats to trade routes, the fact that economic prosperity depends on global stability, and the desire to relieve human suffering, would require a broad toolbox of military capabilities. The white paper carefully avoided the term "expeditionary," as that would remind left-wing politicians of The Netherlands's colonial past, putting the whole restructuring process at jeopardy. The new security situation demanded new missions for the armed forces:

- Out-of-Area crisis management operations in support of The Netherlands foreign and security policy.
- The protection of the home territory and the NATO area against threats resulting from our contribution to crisis management operations.

An innovative idea was to replace threat-based planning by capabilities-based planning. Using the notion of "political ambition" as a main driver, it posited that, as it was now impossible to derive military requirements by examining an existing adversary, politicians had to accept that they had no other choice but to define the number and the nature of contributions to international coalitions they were willing to make.[19] This resulted in the following statement of The Netherlands's defense political ambitions:

- A capacity for a contribution to four peace support operations at the same time with a battalion or equivalent-size unit such as two frigates and a squadron of fighter aircraft. The sustainability requirement was three years, with rotations every six months.
- A contribution to sustained combat operations would require combined arms operations. This entailed a brigade or an equivalent thereof, such as a maritime task force or three squadrons of fighter aircraft.
- Finally, mobilization units would add to the wartime strength of the armed forces of approximately 100,000.

Arbitrarily proposed by the Concept Division, and not underpinned by deep analysis, the idea of political ambitions and its consequences for defense planning was accepted without much political debate or opposition within the ministries of Defense and Foreign Affairs.[20] It was subsequently accepted by Parliament. Political ambitions remained unchanged until 2006, when a new cabinet pressed by budgetary constraints reduced its ambition and started to plan for three simultaneous operations instead of four.

Another breakthrough concerned conscription.[21] Ignoring the advice of an official committee charged with examining the future of conscription,[22] the minister of defense personally made the decision to abolish it. He feared that further downsizing would reduce the intake of conscripts dramatically, and would make the whole system rather unjust for those few still called into service. For political reasons it was also deemed unacceptable to send conscripts into crisis areas against their will. As a professional armed force was more expensive, dramatic cuts in the size of the armed forces were unavoidable. A dramatic downsizing of 30 to 40 percent in terms of manpower would relieve the funds necessary for further restructuring. Moreover, analysis suggested that the armed forces could not maintain its active strength because of labor market dynamics. Reluctantly the army relinquished the 1NL Army Corps, opting in 1995 for the creation of a new bi-national army corps with Germany, headquartered in Münster. The idea of a bi-national army corps was not contested by Dutch politicians and was welcomed by the Germans who were similarly involved in a process of downscaling. The air force saw the number of NATO assigned F-16s falling to 108.[23]

The 1994 *November Brief* (November Letter) by then minister of defense Joris Voorhoeve once again confronted the services with structural financial cutbacks for the following four years amounting to a 6.7 percent reduction of the defense budget. These savings were supposed to be found through intensification of interservice and international cooperation, clustering of common services (such as human resources, medical support, primary training, and maintenance), outsourcing, reduction of overhead, and additional personnel reductions.[24]

By that time the relevance of an expeditionary posture had become evident in missions such as *Deny Flight, Deliberate Force, Allied Force*, SFOR, IFOR, and KFOR,[25] which saw substantial and prolonged Dutch participation. However, these increasingly complex operations challenged the transport and sustainment capabilities and resulted in dramatically increased operating costs. Combined with the ongoing requirement for modernization of weapon systems and an ever-decreasing defense budget, and in addition to the efficiency improvement measures announced in 1994, the reductions announced in the *Hoofdlijnen Notitie* and the *Defensie Nota* of 2000 became inevitable.[26] All mobilization units were now deactivated. The army was to lose half of its main battle tank force, the navy two of its frigates, and the Air Force its reconnaissance squadron as well as its light helicopters.[27] While carefully avoiding the term "expeditionary," the *Defensie Nota* 2000 announced the need for new capabilities such as an amphibious transport ship for the navy, a fully active marines battalion, additional large transport helicopters, and more land forces to improve sustainability.

Looking Abroad, Seeing Shortfalls

By that time, defense policy and planning were increasingly affected by NATO and EU initiatives; at least outwardly that was the impression. After the NATO Defence Capabilities Initiative (DCI) declaration that followed the air operation over Kosovo, and the Franco-British St. Malô European Security and Defense Policy initiative that was followed by the Helsinki Headline Goals, The Netherlands actively supported the objective of improving Europe's military capabilities. The DCI listed fifty-eight shortfalls, divided into areas of deployability, sustainability and logistics, effective engagement, survivability of forces and infrastructure, command and control, and information systems. Six areas of high priority were identified, involving Strategic Lift, Air-to-Air Refueling, Suppression of Enemy Air Defenses, Support Jamming, Precision Guided Munitions, and Secure Communications. This acknowledged the unacceptable European dependency on US military capabilities that had been demonstrated in the skies over Kosovo, in particular in terms of air power assets. The Helsinki Goals mirrored this list of shortfalls. The 2000 White Paper signed by the new defense minister, de Grave, thus emphasized the need for a credible ESDP while at the same time emphasizing the need for a strong NATO.

Satisfying both the national pro-EU stance and the traditional transatlantic orientation, de Grave became remarkably active in the promotion of European defense cooperation projects to enhance national and NATO capabilities.[28] More than just strategic motivations inspired these initiatives. Forced in part by Parliament, the explicit, albeit somewhat idealistic, assumption was also that the creation of multinational units, national role specialization, and international pooling of capabilities, if coordinated in a multilateral context, would result in efficiency savings.[29] With French support, in December 1999 he thus presented a European Multinational Maritime Force concept for ad hoc joint and combined naval operations. With his German counterpart, Rudolf Scharping, and with support of the United Kingdom, he launched the idea of a European Air Transport Command. As a first step, de Grave agreed to spend approximately €50 million on "cofinancing" German strategic airlift capabilities.

In 2001 the EU noted in essence the same shortfalls in its European Capabilities Action Plan (ECAP) after a detailed requirements analysis for the EU Rapid Reaction Force, a paper force of 60,000 troops, 400 aircraft, and 100 naval vessels envisioned primarily for humanitarian operations that could, however, also include some instances of high-intensity combat. In 2002 NATO launched a new but more limited and focused initiative, the Prague Capabilities Initiative (PCC), implicitly accepting the failure of the comprehensive DCI. Even more strongly than before, speeches, letters to Parliament, and policy papers explicitly refer to these initiatives and their underlying problems to inform and justify national policy initiatives.

In the defense budget for 2001 another €100 million were reserved for European cooperation projects. The budget mentioned numerous cooperation projects, including the strengthening of European strategic air and sealift capacities, increased deployability of the German-Netherlands army corps headquarters, Anglo-Dutch cooperation regarding the training of armed helicopter crews, and enhanced European capabilities for air-to-air refueling and combat search-and-rescue. Most initiatives were aimed at improving the expeditionary capabilities of Europe's armed forces. In 2004 The Netherlands supported the concept of EU Battle Groups. In addition the German-Dutch army corps headquarters was turned into a high readiness headquarter for NATO and EU operations. In 2005 a Dutch initiative resulted in the UN Stand-by High Readiness Brigade (Shirbrig).

Money Rules, Again

In 2003 De Grave's successor, Henk Kamp, sent a letter to Parliament announcing additional measures, followed by an update in 2005. Kamp was the first minister who explicitly mentioned the word "expeditionary." His ambition was to play a role in the "premier league." This was expressed in plans for the procurement of cruise missiles for the air defense frigates to contribute to early-entry operations. Kamp also forcefully promoted the Dutch contribution to the stabilization mission in Iraq in 2003, deployed F-16s, Apache and transport helicopters, and a Provincial Reconstruction Team (PRT) to Afghanistan, as well as commandos for high-risk operations to southern Afghanistan to prepare for the deployment of ISAF III. Reconfirming NATO as the main pillar of Dutch security policy, the 2003 plans were underpinned by frequent references to the 2003 Defense Requirements Review (DRR) and the NRF initiative of 2002. The 2003 DRR highlighted both equipment shortfalls and surpluses within NATO, and the Dutch plans should be seen against that background, the letter to Parliament noted. Indeed, the EU and NATO "capability gap" in relation to the expeditionary aspirations, more than ideas such as NCW or EBO, formed the international justification for several substantial procurement plans well into 2006.[30] To close the capability gap, there is a need for transformation among NATO militaries, the 2003 letter observed, and that included the Dutch armed forces. The reductions and cuts therefore needed to be regarded as inspired by two aims: to achieve financial cuts, and to free up funds for transformation.[31]

But again the minister was confronted with budgetary problems. A 5 percent budget cut in 2003 required a reduction in the size of some force elements of 20 to 30 percent. Several frigates and mine hunters, all maritime patrol aircraft, all army reserve units, the army's Multiple Launch Rocket System (MLRS), some artillery pieces, and a number of F-16s were deactivated to finance plans for expeditionary capabilities such as Medium Altitude Long Endurance Unmanned Aerial Vehicles (MALE), extra Chinook transport helicopters, and Hercules transport aircraft. Thus, the 2004 Defense Budget Statement sought a "new bal-

TABLE 6.2

Expenditures by Budget Category

Budget category	Bn Euro
Personnel	3.4
Pensions	1.0
Investments/acquisitions	1.4
Material exploitation/maintenance	1.3
Other	0.2

ance" between the tasks of the armed forces and the budget, in order to create affordable armed forces and the necessary funding for investments.

The "new balance" was a fiction, and some capabilities came dangerously close to losing critical mass. A case in point was the reduction in the number of frigates from fourteen to ten. As a consequence, contributions to peacekeeping operations with frigates were possible only when units were withdrawn from standing commitments—that is, the Standing Naval Forces of NATO or national obligations in the Antilles. Furthermore, the reduced number of aircraft would still require the same infrastructure, logistical base, and training facilities, and could fall below a critical mass in terms of economies of scale. While force modernization was still possible, it was not certain that the funding for transformation was sufficient. Transformation required the government to spend 30 percent of the budget on procurement and research and development. Unfortunately, only 20 to 25 percent was spent on those activities. Despite Kamp's high ambitions, there was a clear risk that The Netherlands would fall to a lower tier in terms of expeditionary capabilities.

Deployments to Iraq and Afghanistan only aggravated these problems. Intense operations resulted in high rates of munitions consumption and increased wear of equipment and a need for replacement sooner than envisioned. Meanwhile, urgent operational requirements also called for extra funds (see Table 6.2 for a breakdown of expenditures).[32] For example, Australian "Bushmasters" costing €1 million each were needed to protect troops against improvised explosive devises. In 2006 that resulted in a decision to postpone investments with a combined worth of around €1.2 billion. In 2007 Kamp's successor, Eimert van Middelkoop, faced the need to take drastic measures once more. As the 2007 White Paper shows, the armed forces were confronted with cuts, including reductions in the numbers of tanks, artillery pieces, and F-16s. Some of Kamp's projects were terminated, such as MALE and the cruise missiles (the decision to terminate cruise missiles was revoked in late 2007).

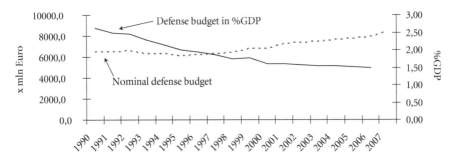

Figure 6.1. Annual Defense Budgets, 1990–2007.

In 2008 this tension was included in the budget statement for 2009. The 2009 Defence Budget actually listed the consequences of prolonged expeditionary operations based on the experiences in Afghanistan. Such operations at strategic distances call for improvements in (and specific additional) logistical and maintenance capabilities, as well as intelligence and force protection capabilities. Moreover, considering the high demand and length of deployments for transport helicopters and aircraft, such operations warrant enhancing capacity in those areas. Finally, the defense budget observes that the high deployment rates of personnel affect the readiness and training of units at home.[33]

The 2007 equipment reductions made clear that, with the Labour Party taking part in the coalition, the government had distinctly less far reaching ambitions. Indeed, manifesting the influence of the Labour Party in the coalition, the coalition of Christian Democrats (CDA), Social Democrats (PvdA), and a small Christian party (Christen Unie) put more emphasis on peace support operations and were reluctant to get involved in high-risk, early-entry operations. In the fall of 2007 the Labour Party published yet another white paper that called for even deeper cuts in the number of major weapon systems.[34] In the context of the debate whether or not to extend the Dutch-led operation in Uruzgan (in Afghanistan), the VVD, the right-wing liberal party, in contrast, called for a halt to the continuous series of cuts of the defense budget that had brought it to well below the NATO standard of 2 percent of GDP (see graph in Figure 6.1). It argued that, while the armed services performed their tasks set by Parliament magnificently, those very same services saw their organizations being hollowed out because of the ever-increasing costs of the operation. The debate concerning the orientation of The Netherlands armed forces was still going on.

FRAGMENTED PLANNING AND CONTROL

The continual series of budget cuts and the reduction of main weapon systems indeed reflected the national debate concerning the nature of security problems and the grounds for legitimate deployment of forces and use of force. In keeping with developments in most parts of the Western world, humanitarian values were considered the prime reason that would lead to future deployments, which were believed to be primarily of a peacekeeping nature. This justified investments in transport and logistics capabilities, and a broadening of the set of capabilities on the one hand. Considering the ever-rising complexity of operations and increasing standards concerning precision, risk-reduction, and responsiveness, emphasizing quality over quantity, as all services did, was also warranted.

On the other hand, lacking an overt direct threat, for the services it also resulted in a problematic zero-sum climate for justifying investments to update or replace existing weapon systems. At no time during the 1990s did the services develop a new national joint perspective on operations or on the future structure and shape of The Netherlands armed forces. Each service had its own budget that was in relative terms distinctly constant over time, with the army budget accounting for about half of the defense budget and the other two services splitting the remainder (the so-called 1–2–1 ratio).[35] Each service valued the peacekeeping operations it participated in as a means for its own institutional survival.[36] No comprehensive and joint debate similar to the RMA discussion in the US took place in which the implications of *Desert Storm* and the IT revolution were explored. Hardly any joint national experimentation was conducted to examine the potential of new joint approaches to exploit data-links, sensors, and precision weapons. All services maintained a very tactical and service-centric perspective on future warfare and operational requirements.

All services also maintained a variety of international or bilateral operational cooperation agreements. The army in part was influenced by its longstanding ties with the German Army, which was reinforced with the creation of the NL/GE Army Corps in 1995. This transmission path concerned training and exercises, command and control procedures, concepts and doctrine, and combined operations experience.[37] The navy emphasized cooperation with the UK and US Navy, and habitually benefited from the political clout that accrued from the intimate connection with Dutch maritime industry, an industry that suffered from strong competition from the Asian shipbuilding industry during

TABLE 6.3

Examples of International Cooperation

Navy	UK/NL Amphibious Force
Navy	Participation in Standing Naval Force Atlantic and Mediterranean
Navy	US Strike Fleet Atlantic
Navy	Drug patrol operations in the Caribbean with French, UK Navy, and US Coast Guard
Army	German-Netherlands Army Corps
Army	CIMIC Group North
Air Force	Integration of AD systems in NATINADS
Air Force	European Air Group
Air Force	European Air Transport Cell
Air Force	Benelux Deployable Air Task Force
Air Force	Extended Air Defence Task Force (with Germany and the US)

most of the 1980s and 1990s.[38] The air force, in turn, with a heavy reliance on US-based training, tactics, doctrine, and weapon systems, often pointed at the evolving US way of war, which hinged upon US air power, as the model for successful joint operations. In fact, in part because such agreements offer a protective anchor for the committed units, the services actively explored additional opportunities for cooperation[39] (see Table 6.3[40]).

Defense planning was thus distinctly disjointed. Indeed, the 1991 *Defensie Nota* had made the service chiefs integrally responsible not only for training and operations but also for personnel, material, and finance, whereas during the 1980s those three areas had been the responsibility of civilian officials at the MOD. The Chief of the Defence Staff had only a very limited capacity to act as "corporate planner."[41] In a budget-constrained environment and with operations continuously ongoing, it is no surprise that within each service, at the policy level, entrenchment, resistance to chance, and adherence to existing roles and force structures were evident. Across the board, only gradual modernization on a weapon system level occurred, and in the adverse budgetary climate the services were engaged in, bitter rivalry sometimes arose when service plans were discussed. Three structural factors fueled this situation. First, there were strong service-specific traditions and cultures. Second, each service operated primarily in an international context, and not a national one, resulting in an absence of any meaningful structural national service cooperation.[42] Third, there was no institutionalized systematic joint operational evaluation—or "lessons learned"—process until well into the 1990s. Instead, if such a process existed at all, it was at service level (but not all services had one), but even then it was marked by methodological flaws and procedural inconsistencies.[43]

The Army

This does not imply that they were "sitting on their hands" operationally, or organizationally. While dealing with the consequences of the abolishment of conscription, the army had to undertake a series of peacekeeping assignments in response to the various crises in the Balkans. While the army was not initially keen to conduct such missions, political pressure, institutional interests, interservice rivalry (the air force and navy were proactive for those very same reasons), and to provide justification for acquisition plans, forced it to concede.[44] This rapid shift from static defense to a high readiness status and a continuous series of deployments, along with new skills required for peacekeeping operations, entailed a revision of training programs and exercises.[45] In response it created a peace operations training center and enhanced its civil-military cooperation (CIMIC) capabilities.[46] Training of combat units emphasized complex humanitarian operations in urban environments. At the same time it was busy making the NL/GE Corps,[47] and the air mobile brigade, a reality, developing the state-of-the-art Integrated Staff Information System (ISIS) for tactical C2, but also improving the postoperations personnel and veteran care system so as to provide specific medical and psychological care.[48]

These new missions and force structures, the end of conscription, the new modes of operations hinted at during operations such as *Desert Storm*, and the new standards concerning casualty and collateral damage avoidance, inspired an audit of the series of army doctrine manuals. Inspired by contemporary British and French doctrinal changes toward maneuverist approaches, with an emphasis on the mental and moral domains of war, the new doctrine was seen as instrumental in helping the army change into one that would be "flexible, mobile, highly responsive and interoperable with allies and partners," the main authors noted in 1995 (the final manual—*Operations against Irregular Opponents*—rolled off the press in 2004).[49]

In response to new missions and new operational areas, it improved the deployability of brigades and battalions, and made efforts to turn them into task force units—the modularity concept—able easily to "plug and play" into multinational formations. Tactical Sperwer Unmanned Aerial Vehicles (UAVs) were acquired to cater for the increasing demands for timely information concerning the immediate tactical environment, and for similar reasons an ISTAR (intelligence, surveillance, target acquisition, and reconnaissance) battalion was created. In part to enhance human intelligence-gathering capabilities, in 1997 the Korps Commando Troepen (the army's commando unit) made the transition into a Special Forces unit. Until the recent operations in Afghanistan, doctrinally it continued to rely on organic assets for delivering firepower, considering airpower only as an emergency close air support (CAS) asset, a mindset harking back to the Cold War. It required an initiative of the air force in tandem with the Korps Commando Troepen to experiment with, and to acquire new equipment for forward air controllers, so as to enhance air-land communications and target designation, thus making CAS missions more effective.

Interestingly, and to the surprise of most defense analysts, the army also managed to get approval for the acquisition of new Pantser-Howitzers, colossal and heavy artillery systems that would require immense transport capacity if they were to play a role in any expeditionary operation. The conceptual and doctrinal rationale behind this was the intention to maintain the capabilities called for by the combined arms concept, which involves the planning and execution of the synergistic deployment of infantry, artillery, air support, and armor. While that capability traditionally resided at the corps level, actual army deployments most often typically involved a battalion and brigade HQ, and the army was therefore gradually introducing combined arms capabilities at brigade level.[50]

A key argument informing the new doctrine, as well as the various changes in force structures and acquisitions, was the idea that an army capable of fighting high-intensity operations could also conduct peacekeeping operations incidentally involving combat actions against irregular opponents such as the factions and warlords in the Balkans. It was a notion similar to the "three block war" concept of the US Marine Corps.[51] In contrast it was assessed that a force optimized for peacekeeping operations would not possess the required set of skills and capabilities to conduct more demanding operations and to shift flexibly between peacekeeping and peace enforcement. There would be no "escalation dominance." In particular, the Srebrenica experience reinforced the belief that the academic boundaries separating humanitarian operations, peacekeeping, counterinsurgency, and limited war were nonexistent in practice. The "blue-helmet" concept had been proven fatally flawed, as the Chief of the Army Staff noted in 1999.[52] While true, this line of reasoning also served the traditional army self-image, of course, for it implied that modern high-tech weapon systems still had relevance and high-intensity combat expertise would still be required.[53]

The Air Force

While going through a series of command restructuring processes and implementing reduction measures, the air force improved the deployability of its squadrons by making them more self-sufficient in terms of logistics and maintenance capabilities. Meanwhile, it maintained a continuous operational detachment in Italy as part of the NATO operations over the Balkans. It merged the Surface-to-Air Missile (SAM) units into one main operating base, while updating the Patriot TBM capabilities with software updates and improved data-links. Helicopter facilities and training were revamped to cater for the new, larger, and more complex platforms such as the Chinook and Cougar, and the new combat role that came with operating alongside the newly established Air Mobile Brigade. Priority, however, went to the F-16 Mid Life Update.[54] This program was considered necessary not only to extend the operational lifespan but also to make a big leap in tactical capabilities (it introduced a much advanced radar, data-links, night capabilities, and precision munitions) and re-establish a proper measure of interoperability and cooperability with US fighters, which had already gone through a series of upgrades.

The air force is an interesting example of the way foreign concepts and technology are adopted for service specific purposes. Contrary to the army, the shift to peacekeeping/enforcements missions did not imply a significant change in terms of doctrine or training, but rather in terms of weapon technology that was required in this era of "precision age warfare." The air force pointed to *Desert Storm* and *Allied Force*, and to NATO Defence Requirements Reviews, as vindication of the enduring—even increasing—value of modern fighter aircraft, which it regarded as the service identity-defining platform, if equipped with modern precision armament, sensors, and data-links. In addition it could point at the role of airborne tanker aircraft and the operational value of the Patriot SAM systems. To that the Apache helicopter was added in the mid-1990s. Not surprisingly, the air force made a deliberate effort to maintain interoperability with US units in terms of technical and training standards as well as in tactics. This strong US influence was (and is) also the result of the frequent training and exercises taking place within the US and alongside US units.[55] Moreover, it was assumed that the US would form the lead nation for most of the future operations in which the air force would take part. An ability to "plug and play" with US forces would of course pay dividends operationally. This view underpinned the investments for updating the F-16 with night-vision capabilities, precision munitions, and data-links. During Operation *Allied Force*, these capabilities had allowed it to play a major role alongside US aircraft, and gained it "A-team" status among high-ranking US generals, which was acknowledged by Minister de Grave.[56] Interestingly, however, this did not extend to UAVs, which were long regarded as technologically and operationally immature, but also as an unwelcome threat to the air force's plans to get political approval for the acquisition of the Joint Strike Fighter as the successor of the F-16. Indeed, gaining political approval for the participation in the development of the JSF, and thus guaranteeing the future of the air force as a modern, highly capable, and independent service, was one of the most important issues to occupy the (fighter-pilot dominated) air force leadership from 1995 to 2002.[57]

The Navy

In addition to the contribution to NATO maritime standing forces, the navy, somewhat stubbornly, adhered to its focus on the Atlantic and cooperation with the US Striking Fleet Atlantic. This was a deliberate policy.[58] It considered itself a medium-size global power projection force and regarded the fleet of frigates the "backbone" of the fleet, as well as the service identity-defining platform (and career-defining). It maintained strong technological linkages with the US, aided by the national maritime and radar industry and R&D establishments. Only gradually did it endorse the increasing need to support littoral and amphibious operations. Still, in the process it was willing to sacrifice its squadron of P-3 Orion Maritime Patrol and anti-submarine warfare aircraft so as to prevent the budget cuts from affecting the number of frigates. The logic underpinning the structure and force mix of the navy increasingly became a subject of political scrutiny. During the 1990s the navy had been engaged in prolonged drug patrol and embargo missions in the Adriatic and Caribbean, employing expensive frigates, although such missions did not require high-tech and heavily armed vessels. In general, littoral missions seemed not to justify continued investments in complex and costly combat vessels that were optimized from "blue-water" operations. It was "creative re-use" of the frigates, but one that The Netherlands armed forces could ill afford, as Labour Party parliamentarian Frans Timmermans noted in 2003.[59]

In response to this changing climate, in consultation with the navy, the industrial maritime lobby group NMC commissioned a study to examine options for creating a proper and affordable mix of surface vessels. It took inspiration from US and UK doctrinal and conceptual developments, including "sea basing," Ship-To-Objective Maneuver, NCW, and EBO. It concluded that the navy needed (1) ships for power projection–related combat operations equipped with ballistic missile defense and long-range land attack capabilities (that is, cruise missiles); and (2) dedicated long-range patrol ships for patrol tasks and humanitarian support. Moreover, to support expeditionary operations on land it was assessed that both a logistics support ship and a Helicopter Support Ship (HSS) were required. With an eye on budgetary realities it assigned highest priority to the acquisition of cruise missiles and the HSS.[60] In essence this implied no substantial changes to the existing fleet, but only additions, without significantly addressing the less demanding patrol tasks. A similar conservative conclusion, justified once again on budgetary grounds, could be read in the official 2004 MOD study concerning large surface vessels for the navy. Here the issue revolved around the appropriate mix of new air defense and command (LCF) frigates,[61] somewhat older M-frigates, and a new type of vessel, the corvette (shorthand for a long-range patrol ship) that would be tailored to conduct drug patrol and embargo missions. The recommendation was to have a mix of four LCF and six M-frigates, and no corvettes. Not surprisingly, Parliament did not concur and tasked the minister of defense to take another look.[62]

GOING JOINT, BY FORCE

Watershed: The Franssen Report

Several developments would gradually improve the disjointed innovation environment. First, after 2000, rising costs of operations, the inability of the Chief of the Defence Staff to effectively control deployments, and a declining budget inspired increasing political scrutiny over the incoherence and inefficiency of the decentralized defense planning process and the operational command and control structure for deployments.[63] Strengthening the role of the Chief of the Defence Staff as chief strategic planner, and creating a Joint Force HQ to centrally initiate and coordinate deployments of Dutch troops and units, were considered promising options to achieving improvements. In 2001 Minister de Grave initiated the Adviescommissie Opperbevelhebberschap (Advisory Committee on the Role of the Supreme Military Commander) with Mr. J. Franssen as chairman.[64] When he presented his report on 19 April 2002, it heralded a watershed in the management of the armed forces.[65] The report proposed a strong top-down defense planning process whereby the Defence Staff would hold the authority to allocate resources for major weapon acquisitions and set priorities in defense investments.

During 2002–3 the MOD took the proposals one step further, also in light of

TABLE 6.4

Planned Expenditures

	2003	2006	2007	2008	2009	2010	2011
Navy	1400	658	604	574	570	569	568
Army	2200	1470	1387	1365	1352	1348	1347
Air Force	1400	673	643	601	601	569	594
DMO		2091	2307	2310	2230	2293	2221
CDC		626	734	681	654	625	667

the budgetary pressures. The services were told to reduce the size of their staffs. The roles of service chiefs would be abolished. Instead, they would become operational commanders with the responsibility to train and prepare their units for deployment. The Director of Operations of the Defence Staff would take control of these units during operations. The service specific plans, requirements, policy, and budgeting capacity would be transferred to the Defence Staff. To improve efficiency all supporting materiel functions would be transferred to one joint Defence Materiel Organization, while all personnel and facilities-related services would be aggregated in a newly established joint Commando Diensten Centra (Service Centers Command). These two new commands would also obtain the appropriate budgets that had resided with the different services up to that point (see Table 6.4).

Observing the US Way of War

Second, starting with Operation *Allied Force*, operational experience too resulted in more "jointness." In particular the recent coercive strategies resulting in regime changes in Kabul and Baghdad markedly affected the debate in The Netherlands, creating a fertile ground for NATO Transformation. Several Dutch studies and articles noted that these operations pointed toward the emergence of a new model of warfare, as well as toward a problem for European forces if they wanted to maintain the ability to operate meaningfully alongside US units. The combination of maneuverable high-tech forces, including Special Operations Forces (SOF), precision-guided munitions, and real-time targeting had proved extremely effective. Networking of forces had contributed to the tempo, which appeared fundamental to the success of military operations. The combination of intensive air strikes with the continuous employment of highly mobile, high-tech ground forces had made it possible for relatively small forces to defeat larger ones. The operations demonstrated that the combination of innovative concepts and the use of small, high-tech forces for advanced expeditionary warfare could achieve objectives in remote areas with astonishing speed, low numbers of friendly casualties, and modest collateral damage. Defense analysts and planners in The Netherlands considered the success of the interventions in Afghanistan and Iraq as an initial validation of the concepts of EBO and NCW. Before these concepts gained a NATO "flavor," however, they were discussed in depth within the services[66] and the MOD.[67] Thus as early as 2000, and in particular in 2001–3, well before such NCW/EBO concepts became commonplace within Europe, key notions of NCW and EBO entered the Dutch military.

The air force, for example, in discussions surrounding the JSF procurement, emphasized those aspects that cohered with ongoing US NCW/EBO developments such as enhanced interoperability, the ability to operate in a "system-of-systems" environment, as well the benefits of a wide variety of onboard sensors for creating situational awareness for air and land forces. Documents detailing additional requirements of the ongoing F-16 Mid-Life Upgrade program likewise play the card of the new model of warfare, including the need for real-time information, and new standards concerning avoidance of collateral damage and fratricide.[68] The navy, too, found no problems in accommodating an idea such as NCW (a concept of maritime origin), as it was already investing in data-links and satellite communications equipment.

Since 2003 these concepts have been the subject of an increasing number of conferences and studies.[69] The *Prinsjesdag Brief* of 2003 makes explicit note of the changes marking the operational environment, such as asymmetric operations, Network Centric Operations, expeditionary operations, joint and combined operations, precision engagement, emerging technologies such as directed energy weapons and nonlethal weapons, and the increased employment of unmanned systems. In speeches, Minister Kamp referred to this new model of joint operations that he also recognized in NATO operations in Afghanistan, explicitly considering it as an integral part of the transformation process the Dutch armed forces were engaged in.[70] Subsequently, NCW-related arguments found their way into operational requirements and policy documents. These came on top of, and dovetailed with, arguments derived from the NATO and EU lists of capability shortfalls.

NCW Embraced, in a Way

NCW itself was given a national twist. Closely following the UK example, Dutch military planners recognized that The Netherlands should aim for Network *Enabled* Operations (NEO), rather than Network *Centric* Warfare. NEO requires units to plug in with their C4ISR assets, while actual operations will be conducted according to national doctrine. Member states should develop the ability to operate in such a network, while NATO and the EU combined should lay out the architecture for such a structure. This would enable units to operate within an EU context and to maintain interoperability in a US-led operation. That holds especially true for land forces, given that navies and air forces already have a high degree of interoperability. In June 2001 the minister of defense decided that The Netherlands should play a leading role in C2 matters in Europe. The aim was to create a European C2 architecture. The initiative was offered to the EU as part of the European Strategic Defense Initiative. As a result, the army's C2 Support Centre spearheaded the development of NEO.

The debate entered a new stage with the 2003 letter to Parliament, which noted that interoperability and the development of Network Enabled Capabilities (NEC) was of vital importance. The NEC concept was developed in close cooperation with the Ministry of Internal Affairs, which is responsible for homeland security. The NEC concept could also be used for national security, because it allows the networking of police and other services to the capacities of the armed forces. For example, unmanned aircraft of the army, Apache helicopters, and F-16s could do reconnaissance flights for the police. It dovetailed with already established plans and procurement processes to invest in C3-systems such as "Link 16" for naval ships and fighter aircraft and *Titaan* to connect the ICT networks of the armed forces. On the ground the Battlefield Management System and the Soldier Modernization Program should enhance the combat efficacy of the individual soldier.

An important milestone was the publication on 11 February 2005 of the MOD NEC Study by the Concepts Division. Pregnant with US NCW terminology and heavily leaning on UK and HQ SACT (Supreme Allied Commander Transformation) NEC documents, this study included a critical analysis of the concept, but recommended the adoption of the overall thrust, rationale, and lexicon of NEC to inform, and to rationalize, future operational plans and requirements. Moreover, it was deemed necessary, based on the relevance of NEC in terms of operational effectiveness and interoperability, to intensify investments in NEC-related capabilities such as ISR, information, and communication capabilities and to make Dutch units "net-ready." While much of the NEC concept required further study, it noted that in principle NEC investments could offer many operational benefits, for instance making possible "dynamic task-groups." In effect, the NEC concept fitted well with the modularly structured Dutch units that would have to "plug and play" into larger coalition or lead nation frameworks. The NEC study was followed by a glossy flyer that explains NEC to the wider public, and includes a statement from the Undersecretary of Defense that "if you're not plugged-in, you're not playing." More important, it was also followed by a series of "action-plans" that lay out in detail the migration paths to creating a net-ready force.[71] In addition, following the success of the army's C2 Support Centre, a C2 Centre of Excellence was established to support SACT in his efforts to transform NATO by providing subject matter expertise on all aspects of the command and control process. The C2 Support Centre focuses on the development of several C2 support systems, such as ISIS, TITAAN, AFSIS, and THEMIS.[72]

When assessed on the scale of the five NATO NEC Maturity Levels that indicate the level of progress in the implementation process of NEC, tangible improvements have been small thus far. In 2008 the prime author of the MOD NEC Study assessed the Dutch Army to be only at level 2 ("deconflict"), and not before 2010 does he expect it to rise to level 3 ("coordinate").[73] On the other hand, NEC has, in recent years, become an increasingly frequent topic of articles in Dutch military journals.[74] Another important indicator of the acceptance of the NEC concept and its perceived importance is the increasing numbers of references to NEC in letters to Parliament to undergird operational requirements.[75] In addition, NEC-specific issues have become the subject of a multiyear research program titled "operating in networks" within the Defence Division of the TNO Research Organization as part of the MOD intention of improving NEC capabilities, the MOD informed Parliament in 2008.[76]

Importantly, amid the cuts announced in the 2007 White Paper, investments in NEC were still deemed essential for a variety of arguments. First, an increasing level of complexity of missions was observed. The Dutch armed forces are deployed across the globe in the most demanding of environments and with the highest possible levels of risk. This puts high demands on logistic support capabilities as well as an ability to operate with a variety of coalition partners in addition to NGOs. Another driver of rising complexity is the fact that Dutch troops will face opponents that employ irregular warfare methods. Such complex missions require a large variety of capabilities and also put large demands on timely observation and on rapid and precise engagement capabilities, as well as an enhanced ability to operate jointly. All this reinforces the need to develop NEC capabilities, the white paper notes,[77] a message that was repeated in another lengthy policy letter to Parliament in 2008.[78]

The introduction to the 2009 Defence Budget, published in September 2008, which explains key policy issues, was particularly specific in this regard. In a lengthy section it announced that to promote security, adaptability, interoperability, and operational effectiveness in 2009, particular attention would be paid to creating a common operational picture, to improving data exchange between command systems, data-networks, and Combat ID equipment, as well as to improving sensor data collection, analysis, and distribution. Over a period of four years up to 2011, it expected that €28 million would thus be spent on improving NEC capabilities, and from 2012 onward €10 million per annum.[79]

EBO, After Practice, Theory Adopted

Absent a strong doctrinal culture and debate on a national level within the MOD, the concept of EBO/EBAO and its derivative Comprehensive Approach were initially incorporated only indirectly. Its gradual formal adoption was facilitated by developments at government level, in particular because of emergent foreign policy approaches, as well as by national operational experience in Afghanistan, which paralleled and was informed by UK and US experiences and doctrinal debates over operations in Iraq and Afghanistan.

During the 1990s and up to 2004, EBO in the narrow sense was regarded primarily (and taught at the Joint Defence College) as an offspring of strategic air power theory and discussed in the context of coercive strategy.[80] Only with the publication of the NEC study did it gain formal recognition within the MOD. The NEC study details the tenets of EBO, arguing that "Effects Based Planning (EBP) forms the heart of Effects Based Operations, and EBP is only possible with NEC."[81] However, the philosophy behind EBO and NATO EBAO (and by extension the Comprehensive Approach) had (although often implicitly) always been the theory behind practice during peacekeeping operations in the Balkans, where stability and reconstruction of necessity required a multipronged yet integrated approach involving military as well as nonmilitary instruments of power and organizations, and a proper mix of physical and psychological effects. Since then CIMIC gained more attention. Furthermore, in 2001 the MOD had launched the IDEA project: Integrated Development of Entrepreneurial Activities, which enabled business specialists to be deployed and assist in CIMIC activities in crisis areas. In 2003, moreover, EBO in the wider sense in the form of the Comprehensive Approach was included in the new Army Doctrinal Publication ADP IIC of 2003, entitled *Combat-Operations against an Irregular Force*, which specifically stated that modern counterinsurgency operations require a comprehensive and integrated political operation of which military operations are just one element.[82]

From 2004 onward, EBO, EBAO, and the Comprehensive Approach also matched emerging national thinking concerning peace support operations, based on operations in Iraq and Afghanistan and humanitarian problems in Africa. Informed in part by debates within the UK and US on counterinsurgency and stabilization doctrine, and in response to the rather volatile context marking those operations, Stabilization and Reconstruction (S&R) succeeded second-generation peacekeeping of the Balkans era as a key concept. S&R is a challenging mix of counterinsurgency operations, peacekeeping, humanitarian aid, reconstruction, and state building, including efforts such as Security Sector Reform (SSR) and Demobilization, De-mining, and Reconstruction (DDR). S&R forces are likely to be deployed in a potentially hostile environment and most likely will have to deal with "spoilers" or insurgents using asymmetrical techniques, such as guerrilla warfare and terrorism. Providing security in such an environment requires counterinsurgency operations as part of the overall S&R effort. However, peacekeeping principles such as impartiality and neutrality still apply. Too much force undermines the principles, thus putting the entire S&R effort at jeopardy. Consequently, force should be used only on a small scale, preferably during covert operations. The limited use of force is a prerequisite for success. Winning the hearts and minds of the population will deprive the insurgents of their base and contributes to the success of a counterinsurgency operation and consequently to stabilization.

This insight was shared also by the ministries of Foreign Affairs and Development Aid. Subsequently, the increasing requirement for such integrated campaigns was the topic of the report *Reconstruction after Conflict*, sent to Parliament in June 2005 by the ministries of Foreign Affairs, Development Aid, and Defence.[83] EBO ideas, and ACT's variant EBAO, were also congruent with the findings of the Interdepartmental Commission "Integrated Foreign Policy," which argue for closer interdepartmental coordination at the strategic level.[84] That year also saw the foundation of the Inter-ministerial Security Sector Reform Team. In 2006 the 3D-approach—Defense, Diplomacy and Development—was endorsed.[85] At the policy level, EBO/EBAO and the Comprehensive Approach thus referred to familiar ideas and reflected national experience; hence in 2006 the idea was incorporated in the first publication of The Netherlands Defence Doctrine (which acknowledged the direct influence of the

British Defence Doctrine), regarding it as one of three concepts that will shape joint operations, along with information operations and NEC.[86]

In 2007 the first major MOD study on EBO was finalized. In it, the EBO philosophy was seen as comparable to the earlier idea of Concerted Planning and Action and to the Comprehensive Approach, both of which by then had met with political approval. Hence, while acknowledging the somewhat immature nature of much EBO literature, the study recommends that the EBO concept has relevance and applicability for the Dutch armed forces.[87] The 2007 Defence White Paper reflected this development. Although not mentioned specifically as such, the EBAO idea surfaces on page 1 when it notes that current security challenges, "more then ever before" call for close cooperation with civilian authorities and an integrated 3D approach:; Diplomacy, Defence and Development. In a number of updates during 2007 the Ministry of Foreign Affairs informed Parliament of its intention to endorse NATO's initiatives in the field of EBAO and the Comprehensive Approach.[88] In the 2009 Defence Budget the intention was affirmed to further strengthen and develop ties with Non Governmental Organizations (NGOs), International Organizations (IOs), and commercial organizations in order to foster an integrated approach during Crisis Response Operations and Security Sector Reform programs in fragile states such as Burundi and the Democratic Republic of Congo.[89]

The increasing acceptance of EBAO and related concepts within the MOD was paralleled by developments within the operational units. Building upon the 2003 doctrine concerning counterinsurgency (ADP IIC) as well as national operational experiences in Iraq and Afghanistan, the MOD experimented with EBO at the tactical level in Uruzgan. A classified list of some twenty-five desired effects was produced. Field teams measured the effect continuously and reported to the operational commanders.[90] Similarly, in 2008 a former commandant of a PRT indicated that he had based his operations on both the ADP IIC as well as NATO's Comprehensive Approach, and in preparing for his mission he had built on the work done since 2006 by the MOD along with the Ministry of Foreign Affairs and of Development Aid in developing a "Second Generation" model of the Comprehensive Approach. In the first generation model those other two ministries were involved in the field activities of the PRT, but the military commander still was the leading agent; the military operations plans were the most developed part of the Comprehensive Approach, and the other two were in the supporting role. During a counterinsurgency conference in 2007, several speakers, including the PRT commandant, also remarked that despite the sound intentions of the Comprehensive Approach doctrine, a great deal remained to be improved in practice, in particular in regard to institutional coordination.[91] In the second generation model the involvement and commitments were more equally divided, and subsequently the action-plans of all three players were developed in more detail.[92] At the Defence Staff, those insights were incorporated in the lengthy Joint Doctrine Bulletin, published in October 2008, concerning PRT operations in Afghanistan. It stressed the need for adhering to a Comprehensive Approach and applying effects-based thinking during planning.[93] It therefore seems that by 2008, acceptance of the EBAO/Comprehensive Approach concept had moved beyond mere passive endorsement.

CONCLUSION

On the face of it, NATO seems quite an influential factor in the Dutch defense community. The series of white papers published by the MOD have been consistent in the central role assigned to NATO in shaping and justifying Netherlands's defense policy. NATO was and is valued as a normative institution in the sense that it offers a prime source of legitimacy for military operations. Institutional efficiency was and is an obvious additional motive for a relatively small nation like The Netherlands. Moreover, most units of the Dutch armed forces have a history of operational integration in NATO command structures and in multilateral cooperation agreements under the aegis of NATO. Dutch units by and large adhere to NATO doctrine, Tactics-Techniques & Procedures (TTPs), and standards. They are assigned to NATO and have participated in NATO operations, and continue to do so. Recent policy papers and requirement statements explicitly refer to key concepts and trends that form the core of the NATO transformation narrative.

The influence of the US too seems quite evident. Building on a strong transatlantic orientation in foreign policy, and considering itself a "loyal ally" of the US,[94] the Dutch armed forces have been open to US military ideas and technology, of which the adoption of the NCW and EBO narrative is only the most recent manifestation. This is further cemented by a procurement history of US weapon systems and by long standing bilateral relations concerning training, exercises, and cooperation agreements. This pertains in particular to the technology-rich Dutch Navy and Air Force, but, more recently, in particular in the field of C2 (also within the army) there is an increase in traffic between the US and The Netherlands. All this suggests that NATO is a strong avenue for the diffusion of American ideas and technology and that NATO is actually quite influential in national level policymaking.

On the other hand, the picture that emerges is that in The Netherlands, until the end of the twentieth century, the change process was driven mainly by internal considerations. While NATO was struggling with its post–Cold War role and was not providing planning guidance for out-of-area operations, within The Netherlands a political and military debate ensued on force restructuring, down-scaling, and force modernization. This preceded the debate on force transformation of the 1999–2007 period. During the 1990s defense planners considered modernization an evolutionary process, whereas transformation contains the idea of revolutionary change. Modernization was concerned mainly with replacing assets and new capabilities. Downsizing and restructuring from a conscript-based to an all-professional armed force took place during the mid-1990s and resulted in an expeditionary posture.

Transformation, driven by technology and doctrine to adapt the armed forces to new operational requirements, entered the debate after Operation *Allied Force*, the 1999 war over Kosovo. For defense planners it implied that force transformation was required to carry out expeditionary military operations quickly and decisively with limited risks to friendly forces and acceptable levels of collateral damage. Conceptually, both EBO and NCW were regarded as icons of this model of operations. Recent deployments in Iraq and Afghanistan have accelerated force transformation with the Stabilization and Reconstruction missions and Counter Insurgency Operations as the focal points.

Indeed, close inspection indicates that the changes that took place from 1990 to 2007 were marked by the following four features, which has implications for theory development concerning the dynamics of military change. First, a set of factors rather unrelated to NATO were dominant in shaping the main trajectory of Dutch defense policy and force restructuring, in particular during the 1990s. A new threat perception (or rather absence of threat), new political priorities, and resulting budget cutbacks have been a pervasive and enduring motive, severely reducing the size of the playing field for the services. Domestic factors and primarily civilian MOD policymakers drove this process, including the reorientation toward expeditionary operations. It confronted the services with a continuous series of imposed radical and often threatening changes in missions, size, structure, composition, organization, accountability, and, in the end, the level of service autonomy, concurrent with the political demands to conduct challenging out-of-area operations.

Second, within the shrinking playing field that was thus created, the driving factors underlying change processes at service level vary among the services. Service identity, institutional interests (legitimacy and survival), military culture, and bureaucratic politics (interservice rivalry) often influenced these processes. At times a service clearly aimed to emulate the sister service of a major country in terms of weapon system development, while on other occasions existing bilateral ties formed transmission paths, and in many cases those two factors were undistinguishable. Experience in operations frequently seemingly lent credence to their policy preferences. And when appropriate the services would point at NATO requirements when they would offer further justification for their own agenda.

Third, identifying dominant factors of influence and transmission pathways is furthermore hampered by another issue. Dutch armed forces always operate as part of a coalition, which is most often a NATO-dominated one. This means that Dutch armed forces have always emphasized interoperability, both in technical terms as well as in doctrinal terms. Since the early 1990s, The Netherlands aims to contribute with "modules": tactical level units and recently modular structured task forces. This coalition setting also implied that The Netherlands Air Force would probably not operate alongside The Netherlands Army but instead would be controlled by NATO commanders in support of a variety of ground units and joint force commanders. The same applies to the navy. The result is an absence of a truly national joint perspective on future operations.

This carries over into the overall character of the innovation process. Whereas the innovations captured in NATO military transformation such as NCW and EBO show benefits in particular at the operational level, and stand for a new model of joint warfare, innovation within the Dutch armed forces has traditionally been mostly of a technical/tactical and shallow nature (versus deep innovation, which includes doctrine and organization) and weapon system oriented, and is the result of a fragmented, incremental, and service-specific process. Service level culture often reflected that of their main international sister services more than that of the other two national partners. No joint military culture existed, a fact that in part inspired the radical restructuring of the Defence Staff and the abolishment of the services. No blueprint, overarching concept, or joint vision document has formed the prime inspiration and logic for unified coherent change.

Fourth, and interestingly for this study in particular, since 1999–2002 there has been a marked increase in the number of NATO concepts actively being debated in journals and conferences, and being employed in policy papers and operational requirements.[95] While the reorientation toward expeditionary operations was mostly the result of an internal process, the adoption of NCW and EBO/EBAO can be seen as resulting from the confluence of external influences combined with norm diffusion, in particular emulation (notably in the case of the air force and navy), the result of operational experiences, and because of alliance interests (against a new threat). This suggests that NATO Transformation is providing a valuable service for the alliance. But here too a caveat is necessary, because the sincerity of adoption is a function of the level to which the transformation agenda coheres with service agendas and MOD-wide budgetary realities and interministerial politics.

A recent study on Dutch defense policy from 1945 to 1990 makes some interesting observations. It notes that The Netherlands government followed most of NATO's recommendations concerning the size and structure of the Dutch armed forces so as to enhance their capabilities and relevance for NATO. Those recommendations directly shaped national plans and policies. NATO pressure also was a prime cause for maintaining defense expenditures at a level of 3 percent of Dutch GNP until the end of the 1980s. However, the study also notes that NATO's recommendations were adhered to only when and as long as they did not necessitate a reordering of national priorities or result in measures and investments that did not fit national budgetary realities or the societal climate.[96] In that respect it seems that within The Netherlands military the dominant drivers of the processes of change are nicely constant.

The Innovation Imperative: Spain's Military Transformation and NEC

Antonio Marquina and Gustavo Díaz

The transformation of the Spanish armed forces has been a long and complex process. It was only following intense debate that Spain chose to join NATO, some thirty-two years after its founding.

Following the death of General Franco in 1975, the direction of Spanish foreign and defense policy was considered an important issue requiring debate among the political parties. An important part of this debate focused on the question of Spanish entry into NATO. However, discussions among the political parties on the values, benefits, interests, and advisability of Spanish entry into NATO faced significant challenges. It was marked by instances of political sectarianism, a profound lack of knowledge of international realities, and a remarkable ignorance of the military implications Spain might face upon entering NATO.[1] It further represented a very divisive topic, even deeper and passionate than the debate on the autonomous regions in Spain. A referendum in 1986, following nearly one decade of debate, solidified this reality: Spain could not come to any consensus on the issue and was not integrated into the NATO military structure.

Although refusing to join the defense organization, Spain signed "coordination arrangements" with NATO. These arrangements agreed to a contribution of Spanish troops that would be under NATO command despite the fact that Spain would not participate in the NATO decision-making process or preplanning of NATO operations.

Despite widespread acceptance and praise by many important senior Spanish military commanders at the time,[2] this "Spanish model" suffered significant shortcomings and had unforeseen consequences. Awareness of this fact, however, took some time and partially explains the gradual approach to military policies taken by the Spanish following the end of the Cold War. The socialist governments in Spain clearly went beyond the clauses approved in the 1986 referendum. Subsequently, once the Popular Party won the general election and took power in November 1996, the Socialist Party accepted, without reservations, Spanish full participation in NATO.[3]

This chapter examines the development and transformation of the Spanish armed forces since Spain's entry into NATO.[4] Its purpose is to allow observations of the continuities in the process despite changes of government. It will also examine and explain several reasons for this continuity, including the lack of serious debate on the issue and an absence of a solid defense culture among Spanish leaders.

THE NORMALIZATION PROCESS

Upon winning power in 1996, the Popular Party sought to normalize Spanish security and defense policies. In an effort to achieve this, the government wanted to ensure a Spanish presence in security and defense organizations, particularly in NATO. Coupled with entry into NATO, the Spanish government further moved toward improving the efficiency of its military forces. The new government also decided to eliminate compulsory military service, which was causing many problems given the high levels of "conscientious objections" to avoid military service, and moved toward the creation of professional armed forces.

Once these objectives were realized, the Popular Party government tried to move forward and modernize the armed forces. At this time, the role of the Spanish armed forces was focused on humanitarian aid, crisis management, and conflict prevention. Since 1989 Spain had participated in twelve UN peacekeeping operations, two OSCE(Organization of Security and Cooperation in Europe) missions, four EU missions, and in the NATO IFOR (Implementation Force) missions (in Bosnia).[5] This required armed forces that could project their military power beyond Spanish borders. The rationale for this change was based on a new strategic scenario, characterized by the lack of a clear and well-defined threat to Spanish territories as well as the emergence of asymmetric risks, crises, and conflicts in different regions of the world that could affect Spanish security.

The Defence White Book published in the year 2000 presented the Spanish defense policy and emphasized the role of the Spanish armed forces in support of external action with a European, Mediterranean, and Atlantic focus.[6] Simultaneously, however, the Spanish strategic plan also presented the defense policy in a "universalistic perception of Spain's world presence" that implied a global defense of Spanish interests and international cooperation in support of peace and stability.

It further emphasized that the complete professionalization of the armed forces had to be achieved "in the year 2002" (when compulsory military service had to be ended) and a process of modernization had to continue.[7] In order to perform effectively, military capabilities, transport capabilities, compatibility, interoperability, and joint action required particular attention.

The armed forces aimed to meet their military needs "by providing them with the best weapons' systems and support equipment available that resources can buy," and the Spanish defense industry was actively encouraged to participate in these initiatives by the government. This resulted in a large proportion of the military projects being joint programs, conceived and managed by the Defence Staff. Of particular interest to this project were high-tech projects for C3I, including the Defence Operations Centre, the Information System of the chief of Defence Staff, the Joint Military Communications System, the System of Military Communications through the HISPASAT (Hispano Satellites) communications satellites, the Joint Intelligence System for Defence, and the SANTIAGO program in the field of electronic warfare. These new services, programs, and acquisitions were necessary to improve the operational capabilities of the armed forces. The goal was to ensure improvements in communications and information gathering, to maintain technological superiority over any potential adversary, and to enhance the ability of the armed forces to project military force outside Spanish territory.

Despite best intentions, not all of the acquisitions matched the proposed guidelines. In-fighting began among the various branches of the armed forces all seeking to modernize their military capabilities. In the end, the Ministry of Defence was forced to play the role of arbiter among the services, providing at least one important acquisition program for each of the branches.

Briefly, the main programs and acquisitions designated for each of the services included the following:

- The army received a combination of tanks and supporting combat vehicles (the Leopard II, Pizarro combat vehicle, and Tiger helicopter). It further received Centauro armored cavalry reconnaissance vehicles, modern helicopter transport capability, new command and control and intelligence programs, and other programs designed to modernize artillery and antiaircraft artillery. (The crucial question of heavy armor transportation outside of Spanish territory was not contemplated.)

- The navy received four F-100 frigates (the crucial question of Spanish antiballistic missile architecture was not contemplated), new amphibious ships and amphibious vehicles, new S-80 submarines, minesweepers, and new helicopters (the navy selected the LAMPS, a sophisticated antisubmarine helicopter, in 2002).
- The air force received a program for an integrated Air Command and Control System and the improvement of communications, the acquisition of eighty-seven Eurofighters, the modernization of the F18s, and the future European transport aircraft (crucial acquisitions for asymmetric warfare were not mentioned).

Coupled with these additions, the 2007 National Defence Directive pointed out the need for a strategic revision of defense.[8] It emphasized the need for joint action, the modernization of weapons and military equipment, and a new organization for defense, required to address new conflicts and concepts of shared security and collective defense.

Curiously, the document emphasized the obligations and requirements that such a reorganization might have on the new common EU Security and Defence Policy. NATO was not mentioned, despite the fact that President José María Aznar was already turning toward it following his growing disillusionment with the EU.

THE TRANSFORMATION OF THE SPANISH ARMED FORCES

In November 2002, a document entitled "Strategic Revision of Defence," which defined the national defense strategy, was made public.[9] It tried to develop the National Defence Directive and emphasized changes that were intended to be a guide for Spanish defense policy "in the mid and long term" (through the year 2015). It presented six basic capabilities that had been defined by NATO following the NATO summit in Washington in 1999. It is important to draw attention to the following points:

First, changes in strategic scenarios reflected those previously identified in the Defence White Book. They included a focus on the following: inevitable innovations in the unavoidable revolution in military affairs; improving critical capabilities and basic requirements for the armed forces such as mobility; and improving the capabilities of projection, sustainability, availability, conflict superiority, protection, and command and control integration. To these initiatives, improvements to ISTAR, modularity, interoperability, joint action, special forces, and net-centric capabilities were also added.

Second, from a technological perspective, it identified technologies that Spain needed to acquire or modernize. These included improvements to reconnaissance and surveillance systems, the compatibility of communications systems, intelligence, and command and control systems. Improvements were also sought for unmanned vehicles that would improve detection, surveillance, identification, and acquisition of targets, as well as an increase in satellite communications capability, and systems that would make possible greater projection, mobility, and interoperability in communications and information systems.

Fourteen further criteria were added for reorganizing the armed forces. In the case of the land forces, specifically the armored and mechanized forces, it was stated that "progress must be made towards a Land Force with significant projection capability." This was to include projection, deployment, tactical mobility, and strike capabilities. The naval forces must "be oriented towards operations in far off littoral theatres, placing special emphasis on the projection of naval power over land." Lastly, the air force "must have all-weather, day/night combat capability, greater transport capability and greater reach and precision."

In general, this period can be characterized as a transformation of the Spanish armed forces that was markedly NATO-oriented. However, justifying these changes, a new EU reorganization of the armed forces was emphasized in order to be able to perform the Petersberg tasks.

Several key elements of Network Enabled Capability (NEC) were considered during the process of modernization, in particular C4I. Net-centric capability was highlighted, identifying the importance of remaining interconnected during the latest military campaigns in order to achieve the desired results. In the future, the linking of all the combat elements in real time was considered, but that would require a "veritable fusion of today's information networks." The armed forces saw several important inertias, specifically in military planning and the acquisition of military hardware, yet nowhere was the Effects Based Approach to Operations (EBAO) mentioned.

It must also be mentioned that the implementation of a professional armed forces encountered difficulties. Initially, the plan had been focused on reducing the armed forces from 200,000 officers and soldiers to 48,000 officers and 120,000 soldiers. However, the rapid implementation of this plan contributed to a rapid loss of military personnel. The Ministry of Defence had to work very hard to increase recruitment, achieve a minimum standard of recruit quality, and retain existing personnel. Compensation remained the biggest concern: salaries earned in the armed forces were not comparable to salaries in the private sector. Meeting these challenges required improvisation and flexibility. The model initially designed in the Law 17/99[10] would have meant excessive turnover and lack of stability for the professional forces.[11] It was impossible to accomplish the target year, 2002, for full implementation.

The following year the 2003 Spanish Military Strategy (ESME) was published.[12] This document, written by a joint and interdisciplinary working group spearheaded by EMACON (Estado Mayor Conjunto de la Defensa—the Joint Defence Staff), had undergone several drafts and taken more than four years to complete. ESME considered that the current global environment contains a wide range of emerging threats and instabilities that could directly affect Spain's security. Although it provided some new insights, the ESME essentially reiterated the same interests and risks presented in the Strategic Revision of Defence.

Based on these risks, challenges, and threats, the 2003 ESME defines missions, principles, lines of action, scenarios, and capabilities.[13] The understanding of the uncertainty in the evaluation of the security environment resulted in a move from the traditional approach of planning forces based on clearly identified threats to a new one based on the identification and acquisition of "critical" capabilities seen as needed to address the growing risks, challenges, and threats that the armed forces had to confront.

The basic idea advocated by ESME was that the Spanish armed forces should retain the traditional tasks of defending the nation from military attacks, maintain self-defense capability, and promote deterrence. At the same time, however, it had to recognize that given the current security environment, it was also an increasing priority to focus on scenarios outside Spanish borders, including conflict prevention, crisis management, and the fight against terrorism. All of this was to be integrated and form a part of collective defense and shared security.[14]

In this sense, the ESME allowed a progression toward a twenty-first century transformation with important changes in mentality and attitudes. This created new ways of thinking and different ways to operationalize and balance national defense with international security, considered as a continuum. The armed forces were regarded as an instrument for state foreign action in dealing with nontraditional challenges and risks. In short, the armed forces had to be expeditionary.

With this document, the armed forces chief of staff could establish strategic military objectives and determine how the armed forces might achieve them. This included fixing the responsibilities for strategic military objectives and the lines of strategic action, and defining the military capabilities that the armed forces required. In this regard several points found within the document can be highlighted, such as deployment capability, modularity, and leadership capacity in a multinational operation outside Spanish territory, the reinforcement of command and control, or the protection of the forces.[15]

THE NEW SOCIALIST GOVERNMENT GUIDELINES FOR TRANS-FORMATION

Following the Socialist Party's electoral victory in March 2004, they tried to modify the "excessive" Atlanticist orientation of former prime minister Aznar's government. In the defense domain, the new government approved a new National Defence Directive, 1/2004, that emphasized a European priority and the development of military capabilities within an EU framework. This priority would be combined with a "balanced and strong" transatlantic relationship. The directive established five broad lines of action:

- A consolidation of the role of the armed forces as an important element for state foreign action.
- A dynamic and constant transformation of the armed forces.
- The cooperative fulfillment, together with partners and allies, of obligations in shared security and collective defense.
- Firm and decisive support for an efficient and multilateral system to resolve conflicts.
- The active participation of the Spanish Parliament in debates regarding major defense issues.

Furthermore, guidelines for the development of the defense policy in the national sphere emphasized the following points:[16]

The defense organization was tasked with drafting a constitutional law on national defense and reorganizing the Defence Staff, rationalizing the structure of the Joint Defence Staff, and establishing a joint rapid reaction force.

It was also considered necessary to reform certain aspects of the armed forces, requiring them to formulate a process for achieving that reform. Their model can be best summed up by quoting, in extenso, nine important points:

1. Define the model of the armed forces, their capabilities, personnel strength, the size of the forces, and their support units in order to make them more mobile and flexible. This will help ensure that they are better suited for joint and interoperable missions with the armed forces of our partners and allies.

2. Promote the reform of the armed forces in accordance with the new model by equipping them with technologically advanced capabilities and by structuring them so that they will be given a graduated response capability.

3. Determine the amount of resources, their level of availability, and the personnel strength needed to meet national requirements, as well as the number and scope of operations abroad in which the armed forces are to be able to participate simultaneously.

4. Devise a new and realistic model for the professionalization of the armed forces.

5. Determine the personnel strength of command staff, troops, naval personnel, and reservists.

6. Reform the military profession by adopting a new structure for corps and ranks, with systems for advancement and promotion that serve as an incentive for dedication and professional effort.

7. Improve equipment so as to enhance the operational effectiveness of the armed forces. A balance must be maintained between the acquisition of new equipment and the maintenance of current operational strength.

8. Encourage research, development, and innovation so as to maintain a high level of technology, thereby improving the operational capabilities of the armed forces and promoting the competitiveness of the national defense industry.

9. Maintain during the current legislative period a sustained budgetary increase of an amount no less than the one from between 2003 and 2005, in order to provide a stable economic environment, making it possible to accomplish the reform of the armed forces.

However, de facto, there was a remarkable continuity. The EU crisis that was apparent in the rejection of the European Constitution and the Atlanticist approach of the General Staff of the Armed Forces, particularly represented by the Chief of General Staff, General Felix Sanz Roldán, plus the nomination of Admiral Francisco José Torrente as the new Secretary General of the Ministry of Defence,[17] implied inter alia a firm approach to several US initiatives within a NATO framework. In the case of NEC, the decision to accept and implement this concept was almost exclusively down to the General Staff.

The civilian authorities nominated in the Spanish Ministry of Defence had little knowledge of strategy and transformation. That was aggravated by the nominations following the 2004 general elections. The rationale for the most important nominations in the Ministry of Defence was limited to one's political fidelity to the leader or minister, and the majority of civil officials selected lacked basic knowledge on the crucial topics and dilemmas facing military transformation. The role of Parliament and the defense commissions can also be characterized as irrelevant. Debates on technology changes and the implications they might have on the armed forces were nonexistent. Thus, the civilian input in the process of military transformation was extremely reduced.[18]

This was one fundamental reason that might explain why NATO initiatives and approaches entered so rapidly into the Spanish official domain with very few corrections. One year following the 2004 general elections, the transformation process continued to show similarities with the basic criteria established in the Strategic Defence Review of 2002. However, several changes should be mentioned.

The National Defence Organic Law 5/2005 considered the armed forces to be a single and integrated entity, thus the mission differentiation among the services vanished. The organic and operational structures of the armed forces were clearly differentiated, the latter being the exclusive responsibility of the Chief of Staff of the armed forces. Common action and the integration of military capabilities of the different services were fostered, given the existing inertias.

Previously, two ministerial orders had regulated the process of military planning,[19] and developed the structure of the Defence General Staff.[20] The new structure included the creation of an Operations Command under the authority of the Chief of General Staff. This unit would serve as a responsible body for researching and studying new military organizational concepts and doctrines, and for the introduction of new technologies. It would also look to increase the effectiveness of the Defence Staff in carrying out assigned missions. The Systems of Information and Telecommunications division (CIS Division) within the Joint Staff was created and became responsible for the plan-

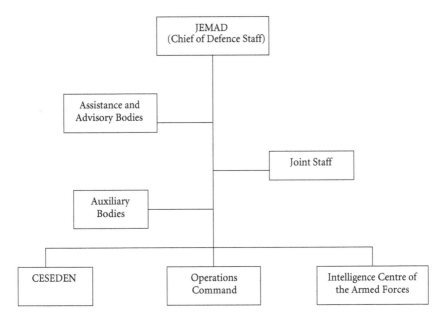

Figure 7.1. Chief of Defence Staff Organization Chart .

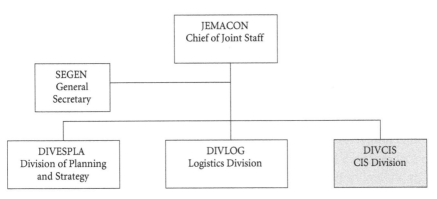

Figure 7.2. Chief of Joint Staff Organization Chart.

ning of the command and control system, and telecommunications within the armed forces.

This new organization (see Figures 7.1 and 7. 2) would allow the Chief of General Staff to facilitate a process of measuring military capabilities and conducting operations. He would be responsible for military planning, and would determine military capabilities and priorities in this domain. Another important aspect to be mentioned is the creation of the Unit for Armed Forces Transformation. This unit is responsible for the research and study of the new concepts for organization and doctrine and its implementation.

According to the Chief of General Staff, the Spanish armed forces had to develop certain capabilities. They needed to be prepared for expeditionary operations outside Spanish territory, and develop joint capabilities for joint and combined action, new joint education and training initiatives, and the integration of the centers for education and joint doctrine. The transformation further required a reduction of between 85,000 and 90,000 troops and 50,000 officers. The main priorities for Spanish land forces included defining the units that could be easily projected, modular and ad hoc for the mission, mobile, effective in combat, and equipped with light weapons systems. Naval priorities focused on the projection of naval power over the littoral and land. For the air force the acquisition of precision weapons for long-distance targets, air transport, unmanned aerial vehicles (UAVs), and, in particular, space systems was underlined.[21]

Royal Decree 416/2006, passed one year later, created the organization and deployment strategies of the Spanish armed forces. By that time, however, it was commonly accepted that all branches of the armed forces had to be prepared to be expeditionary forces. In fact, since 1989 the Spanish armed forces have participated in a total of forty-nine peacekeeping and humanitarian operations (UN, NATO, and EU) beyond Spanish territory with a deployment of nearly 70,000 troops. Despite a socialist government decision to place a maximum number of 3,000 troops on deployable readiness at one time, Spain maintains an active deployment regiment. This restriction will be lifted in 2009.[22] At present, Spain participates in the *Operation Active Endeavour, Charlie Sierra Althea* in Bosnia Herzegovina, ISAF (International Security Assistance Force) in Afghanistan, UNIFIL in Lebanon, and Sierra Kilo-Joint Enterprise in Kosovo. In addition, military observers are deployed in several other countries including the Balkans, Kosovo, Sudan, Ethiopia, Congo, and Indonesia.

Spanish forces also continued to meet their NATO obligations. With regard to the NATO Response Force (NRF), Spain led the land component of NRF 5 (July–December 2005) and has committed to lead the land component of NRF 12 (January–July 2009). It is also currently the leader of the maritime components of NRF 9 and 10 (July 2008–July 2009). The Spanish Air Force continues to participate in the majority of the air forces' rotations, and, taken in sum, Spain is one of the three countries that have made the largest contribution to NRF.

Regarding its commitment to the EU, Spain actively participates in the European Security and Defence Policy (ESDP). It presently is involved in several multinational Euroforces and leads two of the three battle groups with which it is involved.

There are important constraints attached to deploying Spanish forces outside of Spanish borders. Approval by the Council of Ministers, which deployed 1,300 troops of the Plus Ultra Brigade into Iraq, was made without consultation with the Spanish Parliament. This led to a profound debate on the conditions for the deployment of troops outside Spanish territory and resulted in Organic Law 5/2005. This established several conditions on the powers of deployment, including that authorization from Parliament to the government, even in the case of urgency,[23] was required.

Most recently, an important new initiative for military transformation was passed in 2007 that approved the military capabilities goal and consolidated a new system of defense planning. It is now a new, integrated, and flexible system for detecting the needs, rationalization, and optimization of resources, which favors joint solutions for the three services and not partial solutions. The transformation plan addresses 40 military capabilities, includes 185 objectives, and is divided into 7 main areas:

1. The first area supports the state's actions, the surveillance and defense of economic interests such as customs surveillance, and helps the civil authorities face catastrophes, such as illegal migration or terrorism. The creation of a Military Emergencies Unit is oriented to many of these kinds of activities.

2. The second area supports integrated command and control, which clearly backs the use of new technologies and increases surveillance and recognition capacity, with the aim of networking and interconnecting the systems of command and control and intelligence. In this section the capacities for electronic warfare, the protection of information, and the first response to information technology incidents are also included.

3. The third area supports surveillance, reconnaissance, acquisition, and elaboration of intelligence, using UAVs and satellite systems. It maximizes the Armed Forces Intelligence Centre (CIFAS), the participation in multinational programs, and the integration of Spanish intelligence systems into allied systems.

4. The fourth area supports an increase in mobility and projection, with a medium-term aim of improving the strategic capacity of the armed forces. C-295 aircraft would complement the heavy load transportation aircraft A-400M

and tactical transport helicopters NH-90.

5. The fifth area suggests bigger developments in sustainability, the procurement of a new LHD (Landing Ships, Helicopter Dock) supply ship, and the creation of a field hospital. Also in the short term, the support capacity for infrastructure reconstruction, such as landing airfields and support installation, will be improved.

6. The sixth area suggests an enlargement and improvement of the survival and protection of the armed forces, and the promotion of the NBQR (Nuclear, Biological, Chemical, and Radioactive) capacity.

7. The seventh area suggests improvement in engagement superiority, where the previously mentioned Leopard II tanks, the Pizarro infantry vehicles, the Centauro surveillance vehicles, and the new European strike/fighter aircraft Eurofighter, as well as a clear backing for the digitalization of the training systems, are included.[24] Armored units, such as the Leopard II tank and even the Pizarro infantry vehicle, while useful in conventional engagement, are difficult to deploy outside Spanish territories.[25] The international tendency has turned to lighter weapons systems with bigger protection capacity, following the US model.[26]

ENTER EBAO[27]

The EBAO endorsed new approaches for transforming the armed forces. It meant accepting the protocol and looking at military capabilities as a whole. New military planning based on this approach made possible a design whereby the armed forces could work harmoniously, as a set of varying components, working effectively as one in an operational theater. In order to implement this plan it would be necessary to develop new concepts, experiment, and execute military operations based on effects.

The EBAO, which was originally absent in the Strategic Defence Review, was now considered to be the blueprint for transformation because it was impossible to solve military conflicts using exclusively military instruments in the new strategic environment. Successful conflict resolution could be achieved only through the coordinated use of all the instruments of national and international power (economic, political, military, and civilian) and in coordination with other international organizations.[28] This would require new doctrines and training to integrate effectively all forces and agents present in conflict areas.[29] This idea did not contradict the Socialist Party's preference for nonmilitary solutions to conflicts, or for advancing peace. While different actions were undertaken at both the national level and within the EU (the European Defence Agency) and NATO (the Allied Command Transformation), Spain fully participated in the analysis groups created for studying emerging capabilities.[30] The conceptual framework behind the Spanish armed forces' transformation is primarily based on three pillars:

1. Military operations would be carried out based on effects: "*EBAO is the coherent and integral application of the different instruments of allied power which, combined with practical cooperation with NATO's unconnected actors which are present in the crisis area, will create the necessary effects to achieve the posed aims and, as a last resort, the final allied situation.*"[31]

2. Modernization would continue through a system called Concept Development and Experimentation (CD&E), development in the planning process, and the implementation and evaluation of military operations based on effects (EBAO). This led to the creation of the Processing Unit of the Armed Forces (UTRAFAS) as a subsidiary body of the Chief of Defence (JEMAD), which was responsible for researching and studying new concepts and organizational doctrine, as well as those related to the introduction of new technologies. In addition, the armed forces would take into account the lessons learned from recent operations outside Spanish territories, and in certain cases offer solutions to problems it had encountered. This would have to be complemented by intense research initiatives to identify future risks emanating from a continuously shifting strategic environment. The idea of improving equipment, and increasing the operational effectiveness of the armed forces, as well as promoting research and development, are recorded in Articles 7 and 8 of the National Defence Directive of 30 December 2004.[32] It would occur under one of two circumstances. If the capabilities already existed, they would enter into a planning cycle. Should the necessary capabilities be absent, it would proceed to Concept Development and Experimentation, where MIRADO (Material, Infrastructure, Human Resources, Training, Doctrine, and Organization) criteria would be applied. They would be responsible for evaluating all proposals, taking into account the different initiatives, both national and multinational. Although it would be necessary to undertake a new acquisition process to successfully achieve military transformation, it would also have to maintain the present balance between the acquisition of new weapons systems and the sustainability of the present operative forces.

3. Planning would be based on military capabilities, allowing the armed

forces to be designed as a series of systems aimed at achieving a certain effect in the campaign. If, traditionally, operational planning reflected the political objectives of strategic planning at military targets, this new approach would provide a new element of planning. This would allow some changes with respect to traditional procedures. First, in order to plan operations, military commanders would be required to know the desired outcome, which targets would be military ones, and how all of the many forces would be directly applied under the responsibility of the military chain of command or indirectly in support of other governmental agencies or nongovernmental actors.[33]

However, this new model would have to overcome the difficulties of the "pacifist" tendency of civil society, NGOs, and civil ministries. The "culture of peace" and some of the humanitarian approaches practiced by the armed forces under the socialist government collided with the traditional role of the armed forces.[34]

The possible outcomes, however, were unclear. One of the uncertainties surrounded the consequences of applying force in a peace enforcement operation as seen with the Spanish ISAF case. A second unknown was the difficulty that might arise when attempting to work with different organizations, especially those from other cultures. Summing up, the Spanish military transformation had the following plan for achieving the goals outlined in the EBAO:

- Coherence and synergy: exploiting the military operations and activities of allies' potential contributions.
- Deployment and joint sustainability: it should be based on the deployment capacity of the forces when and where they are needed and logistically integrated where possible.
- Superiority in decision-making: it should be based on the information and integration of the network. This allows a decision to be made based on appropriate information and to be put into practice on time against a possible opponent. The culmination of this development process consists of applying information technology to the systems of command and control, which allows the decision-making process to be sped up. In this regard NEC plays a fundamental role.

CIS DIVISION AND NEC

The application of NEC depended primarily on the CIS Division (Systems Information and Telecommunications Division). Its primary mission was to develop the different joint information systems, communication systems, and security of information. This division is responsible for defining the characteristics of the military command and control systems.

Presently, Spain has a military telecommunications system based on radio links and optical fiber that covers its national territory and is complemented by satellite links. Regarding information systems, a system has been developed that will be fully operative by 2012 and complemented by an information security system (INFOSEC). With respect to command and control, there are satellite links through the SPAINSAT and XTAR-EUR that allow Spanish forces full coverage from Denver to Singapore. The principal aim of the CIS Division is to develop NEC, and try to adapt Spanish military command and control to the new NATO philosophy, which advocates effective interoperability and the security of information flow. The Defence Staff participated in a NATO study in 2004 examining its network capability. It was approved in December 2005 by the NC3 Board.

In October 2006, Chief of General Staff Félix Sanz Roldán ordered the creation of a Commission for the Study and Development of NEC that would direct the application process of NEC concepts into the Spanish armed forces. The commission authored a document entitled "Networked Information Concept" that defines its reach and application. The Chief of General Staff's document defines NEC as:

> The capacity to integrate sensors, weapons and posts of Command, both between them as well as with other similar ones (be it civilian, military, national or multinational) at all Command levels (from the strategic level to the tactical one), and which use the same Information and Communication Infrastructure (ICI). By means of its use, information will always be available at any level of decision making, regardless of the place it is and the security guarantees. It provides the Command with superiority in decision making, because it provides it with superior information.

The rationale for developing this concept was based on new risks and threats to national security emerging in the era of globalization. Risks were defined as nonspecific, nonterritorial, and asymmetric, and agents defined as nonstate actors that could have the support of rogue or failed states. These agents could attack the institutions of one country, its communications and infrastructures, its territory, and, in particular, its people. These concerns suggested re-examination was necessary.

In this context, the armed forces would require fewer personnel, more interoperability, and polyvalent forces. The operational command structures of the armed forces had to be interrelated functionally and operationally. They had to achieve information superiority, superiority in knowledge, and superiority in decision and execution. All this implied that a strong effort would be required to incorporate new information technologies, new doctrines, and methods and procedures to share information among all of the agents involved. It was imperative to possess sensors, command and control centers, and interconnected weapons systems plus a networked working organization.

The document, approved by the chief of General Staff, emphasized that NEC needed to impact seven capability areas established in the guidelines for military planning in 2005–6:

1. Integrated command and control. The rapid information exchanges would permit improved mission execution and flexibility for the forces to adapt to situations that might arise midmission.

2. Intelligence, surveillance, reconnaissance, and target acquisition (ISTAR). These will permit increased information diffusion and improved target designation.

3. Engagement superiority. Collaborative planning and information exchange will improve shared awareness of situations and therefore also increase the effectiveness of the joint-combined actions of the armed forces.

4. Mobility and projection capability. The national system's "plug and operate" will allow the armed forces to increase the interoperability among them and their allies.

5. Sustainability. NEC will permit the follow-up of logistics in real time and more secure resupply.

6. Survival and protection. Thanks to the identification capabilities and the network of sensors, the level of risk for the forces can be diminished. The most appropriate weapons system to be used can also be selected in a collaborative way.

7. Government support, thus departing from the parochial support of the Ministry of Defense.

The NEC approach also suggests the use of MIRADO criteria. MIRADO is a set of protocols employed with principles and doctrinal procedures aiming to meet a certain military outcome. The NEC concept also has several characteristics:

- Information superiority;
- Integration and synchronization with other actions in order to achieve EBAO;
- Better capability of response;
- Collaborative Command and Control for dual capabilities;
- Availability of capabilities in the network;
- Information access;
- Information protection;
- Modular design; and
- Optimization of human resources.

Further, the NEC development process is divided into three phases: an initial phase, a consolidation phase, and an implementation phase. A commission, named the NEC Study and Development Commission (CEDENEC), was also created. Initially, the Joint Staff, the Armaments Directorate, and the CIS Division participated in this commission. They were subsequently joined by several other bodies including the following:

a. Planning
b. Communications
c. Architecture and Interoperability
d. Information Systems
e. Security of Information (INFOSEC)
f. Weapons, Identification, and Sensors
g. Doctrine, Training, and Formation
h. Experimentation (R&D)
i. Certification and Validation

The headquarters of the services and the Armaments Directorate participate in the following sections: Planning, Architecture and Interoperability, Communications, Experimentation (R&D), INFOSEC, and Information Systems.

Weapons, Identification and Sensors have not been activated in 2008, given the dispute of leadership and lack of agreement among the services. Certification and Validation was also not activated.

General objectives as well are defined in three phases: short-, medium-, and long-term objectives, whereby interoperability, experimentation, initiative integration, educational models, relations with allies, and operations are contemplated. In the long term, NEC has to be oriented to effects-based operations,

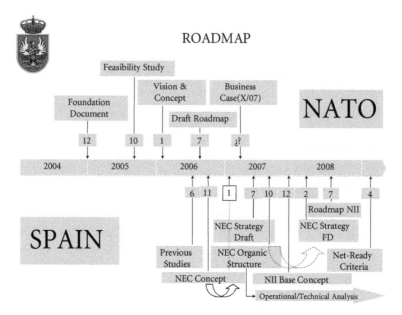

Figure 7.3. Spanish NEC Roadmap. Spanish Joint Staff, CIS Division. 2007.

capabilities must be net ready, and operations must be multinationally networked and include strong civil/military cooperation.

On the technical side, goals are also divided into short-, medium-, and long-term objectives. Long-term goals focus on achieving interoperability, a new communications scenario, full classification, information processing, and an ability to adapt to technical changes.

This roadmap was established by the directorate of CIS, and to date all objectives have been met (see Figure 7.3).

Following an analysis of NEC's impact on armed forces structures, areas affected include the following:

- Security: Security involves not only physical and personnel security but also, importantly, information security.
- Interoperability: The armed forces have made positive steps toward the normalization of systems and infrastructures and in the development of tactics, techniques, and procedures.
- Information Access: The armed forces have implemented a solid Information and Communication Infrastructure (ICI).

- Network and services integration: The armed forces have successfully integrated sensors, command posts, weapons platforms, and support logistics. ICI needs to ensure a sufficient capability to absorb information flows and maintain those connections.
- Standardization and multinational doctrine: The question of doctrine is problematic, taking into consideration that NEC depends on the CIS Division.
- R&D: Concept development and experimentation is considered a fundamental element in the identification, definition, and validation of new ideas, concepts, and the capabilities requested by the services.
- Culture and Education: Emphasis is on a new curriculum and training of the armed forces. The information flow will be so vast that officers and soldiers must be trained for that. It is considered one of the most important challenges for the armed forces.

Until 2008, resources required by JEMAD for the development of NEC have been granted by the government. The parliamentary debates demonstrate relative satisfaction with the resources allotted in the 2006, 2007, and 2008 budgets.[35] JEMAD, which already had a communications system, received funds in the 2007 budget that had been requested for the development of the military information system Bistec1 and the continuation of programs for the development of a new system of command and control. Nevertheless, resources requested for the joint command and control and military communications system were not realized in the 2008 budget. It is critical that resources remain available so that the three services can develop NEC. In 2008, budgetary allocations were promised toward the purchase of CIS systems in the services. However, one challenge which remains is that personnel costs still account for approximately 58 percent of the military budget.

Questions also remain surrounding the political willpower to follow through on these initiatives. Up to now JEMAD has been in charge of dealing with this matter, within the context of the transformation of the armed forces. There have not been debates about the political implications of this transformation process, nor in the parliamentary context or among civil or military experts. However, there remains optimism that NEC will be difficult to stop. Although delays may occur, the support of the three services, the defense industry, and with full Spanish participation in diverse committees and international work groups, especially those within NATO, suggests that some degree of momentum exists. Important tasks remain, however, including developments in the technical field, and, especially, in the doctrinal, educational, and training fields, as well as in the field of experimentation.

PROSPECTS AFTER THE 2008 ECONOMIC CRISIS

In the last few months of 2008, several significant points can be underlined. On 25 November the new minister of defense, Carmen Chacón, presented before the Congress Defence Commission, and later on 15 December before the Council of National Defence, the basic guidelines of the new National Defence Directive.[36] This new directive emphasizes the "systemic integration of all the State resources," the "permanent transformation of the Armed Forces," and the coordination of public and private resources for peace operations and humanitarian aid. The approach is defined as an "integrated approach, inter-ministerial and multidisciplinary."[37]

Later on, in the discourse before the king and the military services on 6 January, the defense minister stated: "The new Directive makes an integral approach to security and defense. It merges military and civilian instruments, public and private, for the defense of our security and our national interests."[38] The emphasis is clear, on joint and interministerial action.

It is apparent that EBAO is still a paramount concept for Spanish transformation, even after the USJFCOM presented several critical arguments against Effects Based Operations.[39] The impact of this critical position on some branches of the Spanish General Staff was remarkable, in particular in the Unit for Armed Forces Transformation. But, surprisingly, EBAO was maintained as a crucial concept. The reason was that the concept was still part of NATO doctrine. The Ministry of Defence has realized that this concept fits well with the Spanish vision of conflicts in the near future.[40] In the coming months it is predictable that some nuances will be introduced in the Spanish approach.

In general we can say that there is continuity with the principal lines contained in the 2004 directive. Another significant point is economic predictability. In order to finish the process of transformation, the new directive points out that political will to maintain a sufficient and continuous budget that can permit a stable economic scenario in the medium and long term must be maintained. However, in 2009 the cuts in the defense budget are noticeable. It is estimated that there will be cuts by 15 percent in the modernization of procurement. The NEC roadmap was slowed down, and it remains to be seen whether the budget for the development of NEC, as originally planned, will be maintained. The general opinion is that the roadmap will be extended to face the economic constrains.[41]

CONCLUSION

Since Spain normalized the NATO situation, the transformation process undertaken by the Spanish armed forces has displayed continuity. The change of government in 2004, despite the Eurocentric inclination of the socialist government, did not substantially derail the policies and ideas that the Popular Party governments had introduced.

This may be explained by several factors. First, there has been a reduced role for political leaders who lack expertise in military strategy and transformation, and they do not have a support staff familiar with these areas. This has resulted in a Ministry of Defence in which the majority of higher posts are occupied by senior officials who have never dealt with these matters. Second, there is the limited role for Parliament. Authentic experts, who are capable of leaving behind common themes and going deeply into technical questions, do not exist in the defense commissions: therefore debates within the Defence Commission lack depth. The political implications of technological matters and developments are not discussed in any capacity. Third, there exists a lack of ambition and paucity of debates about security and defense matters in Spanish civil society and in institutions dependent on the Ministry of Defence. Spain lacks solid and original strategic thought. Adding to the complications are issues including politicization of the issue, attempts by diverse government bodies to supervise and control, the policy of subsidies, and the fomenting from government of a culture of peace. On many occasions, these issues have been perceived as contradicting studies on security and defense.

The result has been the armed forces, and more specifically the General Staff of the Armed Forces, directing a substantial part of transformation, such as the EBAO and NEC, without external input or debate. Military leadership has also provided continuity to transformation processes introduced by different governments with different priorities at different times.

NATO, and debates within NATO, also played a fundamental role in the evolving Spanish defense doctrine. Analysis indicates that where Spain participated with NATO committees and working groups, they inevitably would adopt the resulting doctrine into Spanish policy. It must also be stated that despite Spanish participation in the European Union security and defense policy, EU institutions do not by themselves adequately account for the transformation processes underway in the Spanish armed forces. The role of the Spanish defense industry also plays a minimal role in this transformation process. Most of the equipment and weapons systems used by the Spanish Navy and Air Force have been produced or licensed by US defense firms. This suggests that beneath the surface, American security doctrine may influence some of the branches of the Spanish armed forces.[42]

Additionally, the traditional rivalries within the services have diminished considerably as a result of the changes made, especially by the reinforcement of JEMAD's role and the creation of the Unit for Armed Forces Transformation. It is symptomatic that in the adoption of NEC the role of the services has been practically nil. Because of this there has not even been debate in the military magazines of the services.[43]

As regards the matter of acquisitions, the role of the services has also been drastically reduced as a result of the approval of military capability goals and new military planning based on capabilities.

Spanish strategic culture has suffered because of the absence of an authentic government strategy for the use of its armed forces in foreign policy. This has been most palpable with the socialist government that came to power in 2004. In reaction to the former Aznar government's policy of supporting the military intervention in Iraq, current policy emphasizes a break from this previous position and favors a culture of peace. This prioritizes the armed forces to participate in humanitarian and stabilization operations, and avoids risk-taking. This approach affects the armed forces transformation process and will complicate the application of the EBAO.

Given that a hierarchy of national interests still exists in terms of geographical zones, it appears as if Spain acts as a rear-guard in third-party military deployments abroad. A growing tendency toward multilateralism and international solidarity can also be seen; however, Spanish experiences with southern Mediterranean neighbors, even recently, oblige Spanish authorities to maintain a traditional, merely national, approach to defense problems.

Transformation through Expeditionary Warfare: Military Change in Poland

Olaf Osica

Military transformation had long been viewed in Poland as traditional military change aimed at the reform of a postcommunist defense establishment.[1] This included the restructuring and reduction of military personnel, the modernization of equipment, and the introduction of civilian and democratic controls. Within the last few years, however, the term "transformation" is understood in the context of a "continuous and proactive process of developing and integrating innovative concepts, doctrines and capabilities in order to improve the effectiveness and interoperability of military forces."[2]

The new approach occurred as the Polish armed forces (PAF) have been confronted with an unprecedented number and scale of challenges. On the one hand, lessons learned from troop deployments under NATO and more recently under the auspices of the EU (in the Balkans, Iraq, Afghanistan, and Africa [Congo]) exposed limitations of traditional defense reform. On the other hand, this expeditionary operational experience of recent years seems to exceed the absorption capacity of the state's defense structures. As Peter Podbielski argues, "Poland remains bereft of a strategic review and road maps to guide politicians and defence officials in their planning efforts."[3] The result is a great deal of chaos when attempting to reconcile the long-term vision of military transformation with the development of particular military programs, and with an appropriate procurement policy tailored to the one-year budgetary cycle.

Therefore, it became clear for the Ministry of National Defence (MoND) and the army leadership that the traditional concept of defense reform had outlived its usefulness. Against this backdrop, transformation appears hence as a chance for consolidation of efforts to shape the structure of the armed forces accordingly to its new missions. Despite this positive attitude to the military transformation in Poland, the process appears to be in trouble. The Strategic Defence Review (SDR), the new "National Security Strategy 2007," and a number of draft papers have all suggested that these challenges can be viewed as a new paradigm. It does, however, remain unclear whether the PAF are on the road toward a genuine transformation, or are merely following a simple modernization path with some elements of the new approach. The lack of an appropriate institutional environment (dysfunctional or inertial defense structures) may effectively undermine the ambition of the political leadership to transform the PAF into a powerful instrument of foreign policy within the transatlantic context.

The following analysis of the transformation of the Polish armed forces begins with a brief overview of the strategic and institutional environment in which military change, and in particular transformation, will occur . The first section focuses on key drivers and obstacles to the transformation process. The second section addresses the Polish approach to military transformation, specifically examining the overall concept, expeditionary warfare, effects-based operations, and network-enabled operations. A concluding section closes with a discussion of several factors that will very likely affect the speed and the scale of military transformation in Poland in the future.

BARRIERS AND DRIVERS OF MILITARY CHANGE IN POLAND

There are two sets of problems that have affected military reform in Poland since the end of the Cold War, and continue to have a significant influence upon the military transformation of the PAF. The first aspect refers to factors that impede military change in Poland and include Poland's strategic culture and the unfinished process of overcoming the postcommunist legacy. The second aspect refers to factors facilitating military change in Poland. These include membership in NATO, operational experience from troop deployments, and, more recently, the impact of the EU Common Security and Defence Policy and European Defence Agency as a provider of strategic guidelines and military "know-how."

Transformation Barriers

Poland's current security policy exists in what are arguably its most fortuitous geostrategic circumstances for the past two hundred years. The 2003 Strategy identifies that "the changes in our [Poland's] security environment essentially consist in a shift of emphasis away from the classical risks (armed invasion) that decrease in importance and towards the unconventional risks that originate also with hardly identifiable non-state entities"[4] The 2007 Strategy seems to support this position and declares that Poland is "a secure country"; it places its dependence on energy sources as the most prominent external threat to national security.[5]

Poland has many reasons to be optimistic. It is well embedded in the Euro-Atlantic security context, the threat of a major war in Europe has effectively been removed, and unlike many west European states and the US, Polish territory appears relatively free from the threat of international terrorism. However, one cannot exclude the possibility of retaliatory acts as a consequence of NATO- or EU-led stabilization and peacekeeping operations. Any terrorist operations being carried out against Polish entities on missions outside of Poland fall under the responsibility of the Polish military contingents.[6]

Such an outlook suggests a relatively relaxed approach to military change. The modern armed forces have undergone a shift, becoming more a tool of Poland's foreign policy than a defender of its national independence and territorial integrity. Military transformation should hence be viewed through the prism of political ambitions and financial resources. That, however, is not the case. The problem comes from the proposition underlying the concept of military transformation, which introduces capabilities-based defense planning instead of a threat-based approach. The US Quadrennial Defense Review (QDR) of 2001 reads as follows:

> A central objective of the review was to shift the basis of defense planning from a "threat-based" model that has dominated thinking in the past to a "capabilities-based" model for the future. This capabilities-based model focuses more on how an adversary might fight rather than specifically who the adversary might be or where a war might occur.[7]

This conclusion has long-term strategic implications that are not easy for the Polish to accept. Since the end of the Cold War the mission of the PAF has always reflected the struggle between an "objectively" benign security environment and the Polish strategic culture shaped by memories of a turbulent past (between the late eighteenth century and the end of the Cold War, Poland enjoyed independence only in the years 1918–39). Even as a NATO and EU member, Poland perceives itself as located at the strategic periphery of Europe, and exposed to Russian neoimperial designs. Hostile Russian policies toward the Baltic states and especially the August 2008 war with Georgia acted as a bitter reminder of that fragile geostrategic position. The tension resulting from this mindset keeps the political leadership opposed to a profound and systemic transformation of PAF. Even though Poland formally adopts allied decisions and military guidelines, national defense planning is still based upon the least probable scenario of a military conflict with Russia, and hence prepares the army to fight conflicts of a bygone era. Poland also maintains a separate command system for peace and wartime.[8] Therefore, *quantity* of troops is still believed to be a crucial asset, whereas the QDR approach underlines the primacy of force *quality*. The practical implications of these two attitudes are tremendous if one considers a tight defense budget. Therefore the switch from a threat-based model to a capabilities-based model is a long process that is dependent on identity change, rather than a rational calculus.

The second problem that strongly impinges upon any transformation is the unfinished process of overcoming remnants of the Warsaw Pact, and the persistent legacy of dysfunctional institutional structures—for example, the system of acquisitions and logistics. The structure of the army staff—132 generals, 22,630 officers, 41,860 NCOs, 12,870 privates, and 49,400 conscripts (altogether 127,000 troops)—still resembles that of the "armed forces of old."[9] The final withdrawal of the Soviet equipment that remains a large share of the military capability of the PAF will begin only in 2012.[10]

Finally, Poland lacks a sufficient number of civilian military experts in its Ministry of National Defence (MoND) and the parliamentary commission for defense, a situation that does not contribute to effective cooperation between the appropriate bodies. Therefore, the reform of military training programs and education, including the National Defence Academy, lags behind. Other deficiencies include a lack of viable and integrated logistics, information, and personnel management systems; ineffective and nonintegrated command and control (C2) and command and control, communications, computer, intelligence, surveillance, and reconnaissance (C4ISR) systems; and inadequate research and development programs.[11]

Transformation Drivers

Despite the aforementioned problems, reforms that have been abandoned or have stalled, and a chronic negligence of strategic planning and force development, the PAF have made substantial progress in adapting themselves to NATO standards and the requirements of modern warfare. The pursuit of

NATO membership and the desire to have a relevant army were instrumental in driving Polish ambitions. This was coupled with fears by subsequent Polish governments that a weak army would be equated with "second-category" membership and threaten NATO security guarantees for Poland.

Years of unrealized reforms were partially the result of unrealistic financial estimates and a highly partisan environment in which new defense ministers continued to have "better ideas" than their predecessors. Finally, a real breakthrough came in 2003 with the decision to support the US intervention in Iraq. Operational experience gained in Iraq showed the limits of the traditional reform model, and exposed the real quality of the PAF. As the Chief of General Staff, General Gagor, argued:

> [T]hanks to operational engagements, the armed forces have become a real armed force. Soldiers see the task, understand what the objective of their training is, while commanders get to know what to improve in preparation for crisis response and high intensity operations.[12]

The post-9/11 security environment was best summed up by former US defense secretary Rumsfeld's famous articulation of a "coalition of the willing and able." Poland, as it happens, was politically willing but militarily unprepared to deploy its forces to the Middle East and then assume leadership of a multinational division. However, with the logistical support of the US and NATO, and in part because of unprecedented mobilization at home, Warsaw successfully measured up to the challenge.

Apart from the special force unit GROM, which deployed to Iraq along with the US troops in the first phase of war, a contingent of 2,500 troops (reduced to 900 in 2006) was dispatched to Iraq in autumn 2003. Poland was tasked with peacekeeping and reconstruction in the central-south Iraqi provinces of Babil, Karbala, Wasit, Najaf, and Quadisiyah. Poland also took command of a multinational division composed of more than twenty nations. These years of military presence in Iraq were invaluable for the Polish forces. The total number of troops trained for overseas operations has reached approximately 60,000 to 75,000.[13]

Yet, in order to maintain this momentum and institutionalize the lessons gained from the Iraq experience, Poland clearly needs cooperation from NATO and the EU. The alliance's role is crucial as a body that sets standards and guidelines for troop developments, and assists in the implementation of various programs. Here, Prague Commitments Capabilities (PCC) and NATO's Allied

Command Transformation (ACT) have a role as main transformation drivers. Equally important is US support. Polish-American cooperation has been focused mainly on operational activities in Iraq and Afghanistan, where Poland is using American logistics and transportation. US assistance was also instrumental in establishing GROM, the first special force unit that Poland developed after 1989. In 2007, thanks to extensive cooperation with ACT and US Joint Forces Command (JFC), Poland managed to integrate all indigenous command-and-control systems at the operational command level of the PAF.[14]

The strong Atlantic focus of Poland's foreign and security policy has for a long time been a source of tension with some EU-member states, especially the French. The European defense policy was seen by Warsaw as a potential threat to NATO, or an illusionary project with little or no military application. That position shifted the moment that Poland joined the EU, and the once firm skepticism toward European defense has been replaced by a constructive approach (hence Poland has become a member of the European Defence Agency [EDA]). The "EU Long Term Vision of Military Capabilities," published by the EDA in March 2007 (together with European Headline Goal 2010), is regarded in Warsaw as one guideline for military transformation of the PAF. Poland also became the lead nation in a Battle Group comprising the Germans, Slovakians, Lithuanians, and Latvians. The impact of the EU was also crucial within the PAF for the development of military police (MP) for peacekeeping missions. The MP troops were deployed to Congo Kinshasa in 2006 along with French and Spanish gendarmerie. In 2008, 400 troops were dispatched to EUFOR CHAD/CAR.

TRANSFORMATION—A POLISH APPROACH

The unprecedented institutional, military, and political challenges arising from the Iraq mission were heightened by the decision to increase troop contributions to the International Security Assistance Force (ISAF) in Afghanistan. The number grew from around 100 troops in 2004 to almost 1,200 in 2005, and 1,600 in 2008. These factors had a powerful impact upon the discourse about military change in Poland.

Prior to 9/11 the Polish strategy for meeting the growing demand for troops capable of performing a peacekeeping role was a "two-tier" army model. The security strategy from 2003 advocated that High Readiness Forces (HRF) should be fully interoperable with allies, and have significant power projection capabilities. The Low Readiness Forces (LRF) were to remain responsible solely for territorial defense. Initially, this division of responsibilities between HRF and LRF appeared logical, as it mirrored the difference between the so-called Article 5 and non–Article 5 missions. However, this division was more readily attributed to the poor condition of the equipment and training of the majority of the troops. The two-tier army model in this case led to the de facto continuous deterioration of the troops submitted to the LRF, as limited financial resources were fully transferred to the combat units. This became clear during the Iraq mission when, attempting to contribute a previously unprecedented number of troops, HRF appeared too small to provide the necessary personnel and capabilities, and LRF could barely fill the gap. In addition, the HRF (which were supposed to become the avant-garde of PAF transformation) suffered from the lack of a long-term development vision. Therefore, even though new acquisitions—for example, CASA light-transport aircraft, frigates, self-propelled howitzers, F-16 fighter aircraft, the armored modular vehicle Patria, or the precision guided armor-piercing bullet "Spike"—increased the military potential of the troops, these acquisitions failed to catalyze an added-value for the entire armed forces.

Military transformation appeared on the Polish agenda out of a fear of being sidelined by the rest of her allies. The growing gap between the sluggish reform of the army on the one hand, and the commitments to NATO transformation and requirements of ongoing operations on the other, had reinforced the distance from NATO partners. The strategic consequences of such a development were easy to envisage: the armed forces would lose interoperability and combat-effectiveness, and Poland's influence within the transatlantic context would be drastically diminished.

Strategic Defence Review—A Vehicle for Transformation?

In 2004 a Strategic Defence Review (SDR) team was established within the MoND tasked with "a bottom-up and top-down examination of the defence establishment and the armed forces, as well as preparation of road maps for the defence minister to use in transforming the military by 2020."[15] Although the idea of the SDR was welcomed and had the support of all political parties, it remained politically sensitive for several reasons.

First, the SDR was designed as not only a blueprint for army transformation but also a stocktaking activity as well. This meant that its recommendations, while giving a new impetus to the reform of the most capable services, could also execute a "death sentence" for the structures that effectively resisted transformation pressures. Secondly, the SDR was commissioned at a time when there was no political agreement over the issue of professionalization (that is, its depth and time-frame), the scale of troop reductions (with regard to financial versus geostrategic constraints), and the role of the expeditionary forces in the army. A third problem resulted from the limited scope of the SDR. The reviewing team—composed of six persons—was tasked with reviewing only sections of defense that fell within the competences of the MoND, and left aside the industrial, economic, and demographic realms of defense policy.

The work on the SDR was completed in 2006 following eighteen months of analysis, and submitted to the new defense minister, Radek Sikorski. The review remained classified, but its key findings and recommendations were reported in several media reports. This review, the first of its kind in the history of the armed forces and the MoND, presented an innovative approach to the problem of army transformation: it was driven by a capabilities-based approach. Although the authors of the review were sympathetic to deep, structural changes and the emphasis on troop quality over quantity, the review contained alternative solutions, and put them into a broader financial and geopolitical context. Key questions included the following:

- Whether to preserve obsolete technology, with its coexisting logistical and maintenance backlog;
- How to increase the percentage of NATO standards achieved; and
- Finding the optimum deployable force that Poland can afford.[16]

A lack of access to the actual SDR document prevented the author from fully answering all of these questions. Yet, based upon the general direction of the review findings and a number of confidential interviews, it can be argued that the SDR takes a very pragmatic approach and focuses upon the financial costs of alternative solutions.

One of the key problems that the SDR team confronted was the professionalization of the army—specifically, the pace of the process and the number of volunteer forces in the PAF. When the SDR was launched, the idea of an all-volunteer force (AVF) was still regarded as politically sensitive because of the financial costs and the need to maintain a troop reserve for geostrategic reasons.[17] A mixed model with a growing percentage of professional soldiers was seen as the best choice. This opinion, however, quickly changed within two to three years and moved toward a suspension of the conscription policy, led by the navy and the air force. First, as the army began to purchase new, technologically advanced defense systems, it needed professionals to operate them. Second, the growing number of expeditionary operations and troop rotations increased the demand for volunteers in the army corps. Third, if lessons learned from military deployments were to be institutionalized, soldiers with relevant experience needed to be kept in the armed forces. For all of these reasons conscription became irrelevant or counterproductive. However, the policy advocating a move toward professionalization appeared on the agenda at a time when there was still no agreement on the ultimate size of the PAF. Subsequent governments and presidents stuck to 150,000 troops, and the 2007 security strategy read as follows:

> The size of the Armed Forces of the Republic of Poland will not be altered significantly in the nearest future. Reductions made during the past 20 years or so have brought the size of the armed forces to a level on which the continuation of this tendency may carry unwanted risk. However, armed forces will gradually become professional.[18]

The SDR, however, faced certain challenges. In order to create a sizable AVF, it would require the doubling of defense expenditures from PLN 18 billion to almost PLN 36 billion. This plan would require a financial effort that the state would be unable to afford without significant cuts in other areas, which would be unlikely to be accepted by the public or pursued by politicians. An alternative option put forward by the SDR was a reduction of troop numbers to the level of approximately 100,000.[19] Although these numbers were not accepted by the political leadership, they seemed to reflect what was financially affordable and feasible in light of the growing problems with recruitment. Rapid growth of the economy, in part the result of EU membership and access to its labor market, coupled with increasingly dangerous deployments to Iraq and Afghanistan, made an army career less and less attractive. An insistence on modernization and the negligence of human factors has led to a situation today in which the army has more financial resources to purchase weaponry than to hire professionals to operate the systems. For that reason the SDR proposed to prolong the period of professionalization to nine to ten years, and to maintain the enrollment level of new recruits in an effort to secure troop reserves.

Following prolonged deliberations, a decision was made to adopt an intermediate approach between strategic needs and SDR recommendations. Conscription would not be abandoned but suspended. The PAF would be composed of 120,000 troops, of which 90,000 would be full-time professionals and 30,000 would be commissioned volunteers. In addition, a 30,000-troop national guard would be established.[20]

In March 2006 political leadership of the MoND changed. Minister Sikorski stepped down and was replaced by Aleksander Szczyglo, who refused to accept the SDR. Press reports suggested that the main reason was political. The SDR was commissioned by the postcommunist government of Marek Belka and therefore could not be accepted by a government led by the national-conservative "Law and Justice" party (*Prawo i Sprawiedliwosc*) that underlined its anticommunist vocation. However, an opposite view stressed that the SDR simply "does not reflect the realities of Poland's armed forces and hence is of little usage."[21]

Transformation—Concept Papers and New Security Strategy

From the Polish vantage point the main challenge facing military transformation, apart from the financial constraints and a nascent institutional environment, is defining the military's role in the process. Poland does not have global interests, nor does it feel directly threatened by global security threats such as international terrorism or WMD proliferation. As part of the Euro-Atlantic institutional network, and as an EU member, Poland is, however, dependent on global economic developments. Hence the Polish approach to military transformation appears to be an assessment of its ability to act as a reliable partner in military coalitions, rather than a reflection of national defense needs. Such a departure point determines the scope and the pace of transformation. Clearly, Poland does not need, nor can it afford, to emulate the US approach. Like most NATO members it approaches the US model as an ideal, and seeks to draw lessons and implement solutions that this model offers.

In January 2007 the National Security Bureau organized a high-level panel dedicated to the "Visions of Polish Armed Forces." Based on speeches made during the unclassified portion of the conference, one could argue that Poland anticipates transformation as necessary for the preservation and development of genuine combat capabilities. In the words of the Chief-of-General Staff, General Gagor:

> [T]he technical and technological superiority will remain one of the key criteria of the defence strategy of rich nations. Our approach to modern armed forces will be authorized only if we join the club of nations capable of adapting to critical technological solutions. We live in a time when the firepower does not stand for military strength any longer, and is replaced by power of information.

[Against this backdrop] transformation of the armed forces is an open-ended process; it's a direction that should be followed by not only our, but future generations as well. We need to shape [change], rather than react, rather than adapt, and maximize our own potential of influencing processes and phenomena.[22]

The gap between acknowledging what needs to be done in order to meet the transformation agenda, and what needs to be done to ensure it occurs, however, is large. The conference indicated that among the key structural problems of development and modernization in the armed forces was the need for better interagency cooperation. Further, it required the introduction of more coherence and consolidation into various plans and documents pertaining to army development, including the concept of a command and control system in wartime and military education. The conference also recommended the inclusion of "some" SDR recommendations into the process of force development and planning.

The 2007 National Security Strategy states that the PAF "will achieve and maintain modern army standards by their ongoing transformation directed by assessments and recommendations of military reviews carried out within the framework of the Strategic Security Review." This transformation

> will involve primarily the necessary replacement of armaments and equipment and reorganization of structures to increase operational readiness and troop mobility. The transformation of Polish armed forces will also lead to a more flexible and internally integrated command system in order to ensure its effective operation in time of peace, crisis and war.[23]

A working paper, "Nowa jakość sił zbrojnych" ["New Quality of Armed Forces"], prepared by the president's National Security Bureau (NSB) in November 2007, omits the term "transformation" but instead refers to the notion of a "new quality" that has to be achieved because of three main factors:

- The change in the character of out-of-area commitments of the PAF;
- The change in the art of warfare (resulting from both the type of modern conflicts and technological progress);
- The expanding spectrum of security threats (which may require the potential use of armed forces).

It underlines the need for a thorough analysis of the lessons learned by other NATO members in order to "avoid a repetition of errors made during the process of transformation of their armed forces."[24] It also indicates that "the implementation of 'new quality' through a single, even if complex activity is impossible and that it is unrealistic to simultaneously introduce a process of parallel qualitative and quantitative changes in all functional departments, services and branches of PAF."[25] Therefore, the NSB postulates the creation of "new combined task forces," capable of acting in the changing context of the modern security environment. The "new quality of PAF" would be achieved through the establishment of an "Experimental Task Force" (ETF—Eksperymentalne Siły Zadaniowe) that, when implemented, would act as the nucleus of the new quality approach. Following certification, ETF would then become a model solution for the rest of the PAF.[26]

There is also a second approach, entitled "The vision of the armed forces—2030"—put forward by the MoND's Transformation Department in July 2007. The authors of this approach propose to "conduct changes in *all* functional departments and with regard to the whole armed forces in parallel."[27] The 2030 vision emphasizes that the future of the PAF and its operational capabilities will depend on the visions of the future operational environment from an assessment of anticipated challenges and contingencies. From this perspective, the lessons learned from the expeditionary operations performed by the PAF in Iraq and Afghanistan should become one of the key criteria in assessing the usefulness of adopted solutions in all areas of activities tested there.[28] It suggests a gradual buildup of fully formed units of a modular structure consisting of organic, tactical, support, and logistical subunits.[29] This pattern of development of PAF should enable swift troop rotations without undermining their military preparedness and subsequent combat effectiveness.

There is a consensus among experts and politicians alike that transformation of the PAF has to be considered against a balance between the desire for deep transition, and the necessity of maintaining capabilities to meet current operational demands. Hence, the Transformation Department pledged the establishment of a special fund for the regeneration of equipment intensively used during out-of-area deployments.

The aforementioned two documents pertaining to transformation of the PAF reflected an opposite logic to the process. The "New Quality" approach proposed to start with a small Experimental Task Force, and the "Vision 2030" preferred a holistic approach—a simultaneous transformation of the whole PAF. The first proposal seems more realistic but less ambitious. It takes into account the tight defense budget. At the same time it approaches transformation as more of an experiment than a necessity. "Vision 2030" in turn sets a very ambitious agenda, which might be unfeasible under Polish circumstances. Yet it takes as a departure point the requirements of a modern battlefield, not the domestic political context in Poland.

Expeditionary Warfare

During World War II, Polish troops fought in many remote places around the globe, side by side with Allied forces, in order to secure the postwar independence of Poland. Decisions taken in Yalta and Potsdam that allowed Stalin's Russia to maintain control over Central and Eastern Europe deprived those military efforts of any political meaning. This bitter-sweet memory of military victory and political defeat still colors public debate on military deployments under NATO and EU.

Before 1989, communist authorities developed a tradition of participation in UN peacekeeping missions to underline Poland's "autonomous" foreign policy within the Soviet camp. Several examples include the 1973 Polish military contingent, which has monitored the Israeli-Syrian agreement in the Golan Heights (UNDOF), as well as the small contingent of troops kept in Lebanon (UNIFIL). The 1990s brought new commitments under NATO's aegis in Bosnia (IFOR/SFOR and EUFOR) and Kosovo (KFOR). The latest commitments have been Iraq, Afghanistan (*Enduring Freedom* and ISAF), and EUFOR in Congo.

The nature of these missions varied, but all had some impact upon military change. Poland's contribution to peacekeeping in the 1990s was basically seen as a credibility test with regard to its NATO membership. Missions in Iraq and Afghanistan combine elements of peacekeeping activities with regular warfare and have become important drivers for military change in Poland.

They have helped Poland realize several important facts. First, existing army services were not capable of generating the desired number of high-quality troops; second, the requirements of modern warfare make "the traditional divisions between key services of the armed force disappear"[30]; and third, long-term financial and political (public opinion) costs of the maintenance and rotation of huge military contingents are high. Hence, following years of deliberations, a decision was taken to create a new military service—Special Forces—together with its own command. The security strategy of 2007 reads as follows:

> Special Forces have gained more significance because they are best prepared to carry out operations against asymmetrical threats and to cooperate with other specialized institutions and authorities operating in the state security system. Solutions aimed at effective use of this type of forces should be supported.[31]

For the first time in the history of the PAF's military deployments, the question was posed as to what Poland's political ambitions as a NATO and EU member were, and how they might effectively be matched by the military and financial capacity of the state. This problem had never been seriously discussed before. Previous campaigns—the UN and NATO missions in the Balkans—did not exhaust the human and financial reserves of the armed forces, nor did they require a new command structure. It therefore took Poland four years of NATO membership to decide that the PAF needed a separate Operational Command. The security strategy from 2007 defines PAF's missions in a very broad way:

> The Armed Forces of the Republic of Poland participate in the process of stabilizing the international situation. They are ready to take part in multinational joint stabilization, peacekeeping and humanitarian operations outside Poland's territory ... in operations of asymmetrical nature, including multinational, joint anti-terrorist operations carried out in compliance with international law, organized by NATO, EU or an *ad hoc* coalition of states.[32]

The authors of the aforementioned "Vision 2030" assume that the PAF will be engaged mostly in low-intensity operations but acknowledge a potential for high-intensity missions. According to a representative of the General Staff, the level of political ambition of Poland requires from the PAF a capacity to participate in the following:

- One operation in the framework of collective defense; with a division force with sea and air force component without rotation capacity (16,000 troops); or
- Two high-intensity crisis management operations with the force of an augmented brigade with sea and air force component (up to 5,000 troops each); or
- A number of low-intensity crisis management operations, with a battalion combat force each (2,500).[33]

In January 2009 the liberal-conservative government of Donald Tusk approved a document entitled "Strategy of Participation of PAF in International Operations."[34] The strategy sets the level of 3,200 to 3,800 as an optimal number of troops deployed out of national territory.

The strategy sets political guidelines for future deployments. Based on lessons learned from previous deployments, especially in Iraq, the strategy does not preclude participation in ad-hoc coalitions, but sets a priority for NATO-

and EU-led operations. The document also states that "the scale and geographical areas of deployment ought to be the resultant of the current military capabilities of Polish Armed Forces and clearly defined political ends." The participation in operations is dependent on three rules: the conformity with national interest, latitude of action (troops must maintain the greatest possible influence upon the conduct of an operation), and troop efficiency (an optimal relationship between committed means and declared ends).

Poland will also strive for a better "visibility" of her deployments. This commitment comes from an observation from ISAF, where Poland had a contingent of more than 800 troops operating without national caveats in different provinces. Its "political visibility" was hence diminished. As a result, the consolidation of troops in one province (Ghazi) was viewed as necessary to single out the Polish commitment against that of other allies.

The strategy reiterates the need for full interoperability of deployed troops with armed forces of allies and partners; and declares the intention to include in international deployments not only the army component but also the air force, the navy, special forces, and gendarmerie.

Effects-Based Operations (EBO)

The concept of EBO is not a new idea, but rather a new approach to the old problem of how to "maximize efficiency and minimize wasted effort in the pursuit of goals."[35] It is hence conceived as a guideline for military personnel to operate in a new security environment, in which physical destruction needs to be limited in scope and closely linked with political objectives, which are pursued by different agencies, both military and civilian. EBO are defined as "operations conceived and planned in a systems framework that considers the full range of direct, indirect, and cascading effects—effects that may, with different degrees of probability, be achieved by the application of military, diplomatic, psychological, and economic instruments."[36]

EBO are a totally new experience for the PAF. The already mentioned "New Quality" working paper developed by the National Security Bureau identifies EBO as "an intellectual challenge that needs to be addressed by stakeholders with an aim to adapt future organizational structures of the PAF to EBO's requirements."[37] The paper provides a thorough analysis of the essence of EBO and elaborates on its own concepts in accordance with ideas implemented by allies, members of the EU, and other coalition partners.

At a practical level the Polish MoND (Transformation Department) participates as an observer in two aspects of Multinational Experiment 5 (MNE-5) conducted by ACT: *Effects-Based Approach to Multinational Operations*, and *Effects-Based Assessment*. Scheduled for 2008–10, MNE's key theme is as follows:

> a comprehensive, whole-of-government approach, using the effects-based approach to multinational operations to explore military support to various aspects of civilian interagency operations. By using a combination of national and international elements of power to influence a regional environment, participants will seek to broaden the context of pre-crisis and crisis management. The experiment will attempt to develop processes, organizational structures and/or technologies to support a comprehensive approach in securing a politically stable, economically sound environment by engaging interagency organizations, and increase interagency interaction with non-government organizations.[38]

It remains to be seen how institutionalized the EBO will become in the operational practice of the PAF. One of the systemic problems that must be overcome is a result of the political approach to expeditionary warfare and peacekeeping activities. Although it is the Polish state that is engaged in military operations, from an institutional point of view it usually falls on the MoND to carry the financial and organizational burden of these operations. The interagency approach remains ineffective, as there are no fixed regulations allowing the MoND to "lease" civilian experts from other ministries. There are also no financial means (for example, pay per diem) available for paying experts.

The aforementioned "Strategy of Participation of PAF" from January 2009 underlines the importance of Civil-Military Cooperation (CIMIC) and pledges the creation of a system regulating all legal, financial, and institutional aspects of participation of civilian experts in military operations.

The planned establishment of a PAF Training and Doctrine Centre in Cracow may be instrumental in the creation of a Polish approach to EBO. The center's key task would be to create a cohesive doctrine for force training and development for expeditionary missions. The doctrine would define areas of responsibility and duties of personnel that develop force planning. The center would also become an analytical unit, gathering and processing lessons learned by the PAF from expeditionary operations.[39] A significant role is also played by the Joint Force Training Centre (JFTC) in Bydgoszcz (Poland) that supports training for NATO and partner forces to improve joint and combined tactical interoperability.[40]

Network-Enabled Operations (NEO)

Although the challenge of NEO had long ago been recognized as one of the key preconditions for interoperability with other allies, little was done to make substantial progress to that end. This, however, could not be avoided. Back in the 1990s, the PAF and MoND were struggling with the introduction of basic IT technologies and computer systems for better institutional efficiency in managing services and departments. The lack of a coherent policy, strategy, and procedures, as well as the permanent rotation of staff, including the political leadership, produced a great deal of chaos and inaction. Therefore, computerization of the PAF, the General Staff, and the ministry has often fallen victim to spontaneous, uncoordinated, and sectoral approaches to the problem. Different departments and branches of the armed forces, as well as different types of units, developed and purchased their own platforms. As a result of these practices the MoND today possesses many different IT systems that require tremendous attention and resources to be integrated.

The implementation of technologies that will enable the PAF to operate in a network-centric environment therefore requires the mental and institutional breakthrough that for many years had hampered transformation of the PAF. Today, however, it appears that these changes can now begin to happen. The new dynamics triggered by troop deployments following 9/11 drastically changed the stakeholders' approach to C4ISR; it is no longer viewed as a kind of a "fancy" idea or "eccentric" NATO jargon, but an indispensable capability that the PAF must develop if it wants to operate with allies in a high-intensity conflict. In a speech delivered to the National Security Bureau, the Chief-of-Staff of the PAF, General Gagor, stated that:

> irrespectively of the character of future commitments (in the 15–20 year perspective) Polish Armed Forces ought to possess first of all:
> - An integrated command and control system, which is interoperable with allies and capable of functioning within a NEO environment;
> - Coupled with an integrated system of reconnaissance, troops identification and targeting.[41]

The challenges moving forward are both serious and pressing, as the present command system in the armed forces is based mainly upon the stationary-mobile network of radio-cable contact, which limits the capacity of the armed forces to participate in high-intensity joint sea-land-air operations. In 2008 about twenty-five IT projects were implemented that should form a basis for building up a network capability for NEO. Since 2006, troops have been provided with military satellite systems (VSAT), mobile command post modules, and integrated teleinformatic hubs.[42] A second priority area that needs substantial improvement is "friend-or-foe" identification and the Common Relevant Operating Picture. In the words of the director of the IT department in the Polish MoD, the key challenge of NEO is the departure from a model in which the army command manages every single relation in the chain of command, to the model whereby commanding means distribution of information.[43]

A harbinger of the network-centric era in the PAF is the program "Soldiers of the 21st Century," which in the near future will develop into the program "Polish Army in the 21st Century." The key idea of this program is the creation of combined units that are able to operate in NEO. The program starts with system integration at the strategic level, mainly in the Operational Command, and works its way down to the tactical level, integrating chosen units and equipment of the army's particular services. A future step will be to integrate the strategic with the tactical level.[44]

Perhaps the area in which Poland can make profound progress in NEO in the coming years is the air force. The purchase of forty-six F-16 Block C/D supersonic fighters already confronted the MoND and the air force with a network-centric reality. If the fighters are to be used in any international missions or guard Polish air space effectively, they must be part of a network.

TRANSFORMATION PROSPECTS AND DILEMMAS

The notion of "military transformation" has become an acceptable part of discussions pertaining to the future of the PAF in great part because of the military deployment of significant numbers of troops to Iraq and Afghanistan. However, as this article has shown, Poland is at the beginning of the process. Transformation as a new paradigm is understood comprehensively only by a small number of experts and military personnel. Political elites are still attached to the notion of modernization and restructuring. Military engagements are a key driving force in PAF transformation, although it could also be argued that autonomous, internal pressures for change exist as well. It remains to be seen whether the momentum for change can be maintained. The outlook seems optimistic, but there are a number of challenges that, if badly handled, may slow down or even undermine the process.

First, Poland needs to address the gap between the rhetoric of transformation and the genuine transformation of the armed forces. As the SDR of 2006 was rejected by subsequent political leadership and never discussed in the wider public, MoND and NSC spent time and resources on the development of new transformation documents: "Vision 2030" and "New Quality of Armed Forces." Neither was met with strong political backing, and they remained intellectual exercises instead of guidelines for transformation. A new SDR is scheduled for 2009–10 and probably will have to refer to this problem. Otherwise, transformation becomes a permanent topic for seminars and SDRs without any leverage upon the real functioning of the PAF.

Secondly, after years of economic boom, Poland, like other European countries, faced economic slowdown that might lead to a recession. Such a prospect will have a direct impact upon the pace and scale of transformation. In 2009 conscription was abandoned and Poland set the course for an AVF. That decision does not yet transform the PAF into a professional army; it is barely the first step in that direction. It needs to be accompanied by better training and the acquisition of new equipment. However, the short-term costs of an AVF are high and need to be thoroughly calculated against the financial capacity of the state. Otherwise the PAF run a risk of changing solely the legal status of soldiers whose equipment and training would not improve.[45] Therefore, MoND will have to strike a balance between the need to modernize equipment and the total number of the force, which was set for 120,000. The decision became more difficult when the prime minister, Tusk, demanded that Defence Minister Klich in late January 2009 reduce ministerial expenses of around PLN 4 billion (roughly €1 billion) as part of the national savings package. Tusk also questioned the bill privileging MoND a budget at the level of 1.95 percent of GDP every year. As a result MoND had to abandon its modernization program for 2008–18, and to reconsider the affordable number of troops in the AVF. It remains to be seen whether budgetary restrictions and the threat of abolishing the "1,95 bill" will facilitate a more robust defense reform, tailored to the state's actual needs, or whether it will mark the beginning of the end of transformation, before it actually was launched.

Third, like any other army, PAF functions not only in the operational context of the alliance policy but also in the political context determined by domestic debates and the political objectives of the state. Therefore, in response to a growing resentment among Polish public opinion toward operations in Iraq and Afghanistan, and criticism of the US security agenda, the political ambition to invest heavily in the army's expeditionary capabilities may not be shared by the public. It is therefore regretful that the transformation of the army, including the development of an AVF, is not a subject of public debate. A "white book" on defense that was to be published after the SDR 2006 team completed its work was never published. Because of the public's limited experience with a professional army, there is also little understanding of the various trade-offs involved in accomplishing this goal. Without a new information policy that attracts public attention and interest to issues of defense reform and professionalization, there is little chance for final success. Transformed and professional armed forces need identity that is shared, or at least not rejected, by the citizens.

Finally, the evolution of Polish strategic culture may be halted as a result of neoimperial tendencies in Russia. The August 2008 war in Georgia was a reminder that Russia is still unpredictable and views military force as a political asset in her neighborhood. Should NATO and the EU neglect the eastern dimension of European security because of a full-time commitment to expeditionary operations in the Middle East, Africa, or Asia, Poland may change its defense priorities and refocus on territorial defense. That would be a de facto "death sentence" for transformation, and would undermine the combat effectiveness of PAF, which results from a permanent, on-the-ground interaction with other allies.

Conclusion: The Diffusion of Military Transformation to European Militaries

Terry Terriff and Frans Osinga

The idea of military transformation is widespread in Europe. Even non-NATO states such as Finland and Sweden have embraced the notion of transforming their national military organizations. The idea of military transformation originated in the US, and it has been the American approach to transformation that serves as the original source of military transformation in Europe. For NATO member states, however, a crucial source for their own transformation ideas has been the Alliance itself. At the 2002 Prague Summit the member states of the North Atlantic Treaty Organization agreed that the Alliance needed to transform itself militarily. A key change made to effect this goal was to turn Allied Command Atlantic into Allied Command Transformation (ACT) with a mandate to develop and drive military transformation. Since ACT was created officially in 2003 it has sought to generate new concepts and capabilities to enhance the effectiveness and cohesiveness of NATO's military arm, proselytize for the need for and benefits of military transformation, and help requesting member states develop their own transformation road map. Military transformation in the European NATO member states explored in the comparative cases in this volume is far from complete. Arguably, military transformation can never be finished if one accepts that, by definition, military transformation involves continuous adaptation and innovation. Leaving aside such definitional pedantry, the European militaries examined are engaged in the process of military transformation, in that they appear to be endeavoring to become expeditionary capable, adopt an Effects-Based Approach to Operations (EBAO), and develop Network-Enabled Capabilities (NEC).

Practitioners and analysts alike need to develop the means to understand change in military organizations, to be able to account for why change does or does not occur, and to account for variations as well as similarities across military organizations. The six case studies of European military transformation contribute to the understanding of change in military organizations and also furnish a number of insights of a more practical nature. Thus the specific studies presented in the volume have both policy and theoretical implications.

POLICY IMPLICATIONS

The analyses in this volume identify a number of findings of a practical nature that will be of interest to civilian and military policymakers, both in NATO, including Allied Command Transformation, and in the individual member states of the Alliance.

NATO's commitment to military transformation is aimed in part to stem the growth of the military gap between the US and Europe, both with respect to narrowing the capabilities gap as much as possible and bridging the doctrinal/conceptual gap. What the cases examined indicate, however, is that the implementation by individual member states of their transformation programs is subject to a range of specific national influences, and what appears to be emerging are multiple gaps within NATO Europe itself. Simply put, not only is it not evident that the gap between the US and Europe is being lessened to the degree hoped but also there are both a military capability and a conceptual gap starting to materialize among NATO's European members. There are three factors that provide impetus to the emergence of these gaps within Europe: variable stages of transformation; the impact of domestic factors; and differing interpretations of transformation.

One reason for the emergence of these gaps within NATO Europe is that different European militaries embarked on transformation at different times and so are at different stages in the process of military transformation, of determining and implementing their national policies to become expeditionary capable and incorporating NEC and EBAO. This aspect is clearly illustrated in the case of Poland, which in no small part as a result of its operational commitments as well as the scale of the transformation effort, began in earnest to undertake transformation instead of modernization only late, compared with other states examined. Poland is still very much in the earliest stages of transformation, still working on identifying what it needs to do to become more expeditionary capable and how it should interpret and implement EBAO and NEC. The other five states examined, on the other hand, began their process of transformation earlier and hence are further down the transformation road than is Poland. That a transformation gap can be identified across the six states examined strongly suggests that across twenty-eight member states there will be a range of start points, indeed potentially a significant range of start points, with some member states falling broadly within the spectrum covered in this volume with yet others almost certainly lagging behind Poland.

A second set of factors influencing the development of an intra-European gap is that different national militaries will progress in their transformation effort at different rates, dependent on internal factors. Germany, for its part, was an early adopter of the idea of transformation, but nevertheless looks as though it may start to fall behind, with its transformation effort in early 2009 seeming to be adrift because of the absence of specific political and military leadership willing to continue to drive the process forward. As Borchert's analysis makes clear, this problem is also a function of a domestic governance system that values consensus and a domestic political culture chary of the employment of military force—and hence wary of defense efforts seemingly designed to enhance the Bundeswehr's capability to operate abroad. Spain, for its part, also has made progress implementing its transformation road map, but advancement has been considerably slowed in the past year or so because of growing resource constraints. The case of Spain, as Marquina and Diaz note, further points to the impact that individual services may have on the process, with Spain's individual services in the initial phase of transformation continuing to pursue specific service procurement policies that arguably were inconsistent with the overarching approach to transformation. The cases of Germany and Spain highlight that domestic factors, such as resource availability, strategic culture, political system, political and military leadership, and organizational inertia, will exert an important influence on the rate of progress any one particular European state is able make once it has determined to undertake military transformation. The reality is that each and every European state is subject to this range of internal factors, the combined influence of which means that progress in their individual transformation processes will proceed at different rates, potentially with some advancing by fits and starts, and some failing to make much headway; such differential progress will exacerbate the formation of a military capabilities gap across Europe.

The third factor driving the emergence of gaps within Europe is that each national military organization may be interpreting and conceptualizing military transformation in a different manner. The core of the military transformation process of each state studied in this volume is that of expeditionary-capable force, NEC, and EBAO, but it is far from evident that they interpret these concepts in the same way or are developing them in the same way. There is, then, the growing possibility of the emergence of a conceptual, or doctrinal, gap within Europe as well.

The potential variation in the interpretation and meaning of expeditionary-capable force, NEC, and EBAO arises from the presence of a number of factors. France's official position, for example, is that the purpose of the French military is to "control violence" and conduct peace operations, while the general aim of transformation involves fighting and winning wars. As Rynning suggests, this apparent dichotomy in purpose may inhibit transformation, and by implication may also result in an interpretation of transformation concepts, such as EBAO, that may be subtly or palpably different from the conception and practice of these concepts by other European states. The possibility of a distinctive French approach is heightened by France's ambitions with respect to the European Security and Defence Policy. Germany represents a case of another way in which transformation concepts may diverge. At the time of writing, as Borchert notes, the German Air Force is the military service that has taken the lead in conceptualizing EBAO, leaving the other services to define their particular positions in terms of the Luftwaffe's conceptual approach. Although Germany's efforts to define and implement EBAO (and transformation more generally) are at present moribund, this raises the conjectural possibility that the German military's conceptualization of EBAO may end up more skewed toward an air force–oriented approach than might otherwise occur in a fully integrated joint service or top down generated conceptual development process to which the individual services have to adjust.

That there may be in train possible divergence in interpretation and implementation of the content of transformation is further underscored by the European states' development and refinement of their concept of transformation based in part on their recent and current operational experiences. The lessons to be incorporated into concept development that are derived from the operational experiences of the British and Dutch forces fighting in southern Afghanistan may not be the same as the lessons the Germans, French, and Spanish derive from their operational experiences in Afghanistan. The exact lessons that each derives from its experiences in Afghanistan will be based largely on the national mission sets each nation is conducting there, and the approaches employed by other national military contingents with which they closely conduct operations, with those lessons being conditioned by national strategic and military culture and national aims. In contrast, the French military does not see Afghanistan as being a relevant engine for its transformation efforts. The clear implication is that, to take the example of EBAO, the same concepts may be conceptualized in much the same broad way, but the individual national approaches and the specific factors at play will influence the manner in which they implement and apply the concept, and these various influences will likely impart subtle or even conspicuous differences in content or approaches to practice.

Moreover, the six case studies indicate that European states do not necessarily look to one source to provide the concepts (NEC, EBAO, expeditionary-capable force) that they draw upon for their own use. Some NATO European states do look to Allied Command Transformation (ACT) to provide the model to be emulated. Others, however, seem to look at the range of possible concepts that already exist and then draw on those that appear to suit most closely their specific requirements, and then interpret and develop the chosen conceptual form to fit their particular national interests, domestic political environment (especially domestic budgetary constraints), and internal service interests. What this means is that national interest, budget constraints, and internal military service interests condition which model of NEC, EBAO, and expeditionary-capable force is chosen and, further, how these chosen exemplars are then interpreted and implemented.

European militaries thus may employ a common language of "transformation"—including NEC, EBAO, and expeditionary-capable force—but the actual content, or meaning, of these concepts and how they are applied may be different. Individual militaries may be using the language of transformation more to impart legitimacy to themselves, in terms of how their allied military organizations, or even how NATO, perceives them (or so they can use the language of transformation when talking to them). Or the language of transformation may be employed to sustain domestic defense budgets. European militaries therefore may use the same terms and language, but they may not be talking about the same thing that other allied militaries are, or even NATO is, when using the language of transformation. European militaries may be talking the same talk as other militaries or NATO, but that does not necessarily mean that they are walking the same walk.

As a consequence of the impact of the three factors identified above, there is both a military capabilities gap and a military conceptual gap emerging in

NATO Europe. The degree and extent to which a conceptual gap will emerge is as yet difficult to classify, but the six case studies strongly suggest that such a gap will be present. An additional complication, moreover, is that the military capabilities gap and the military conceptual gap will not be coextensive, which imparts a layered complexity to the very uneven distribution of the extent and trajectory of military transformation within Europe.

This layered complexity of the extent and trajectory of transformation in Europe poses a serious challenge for ACT. ACT recognized very early that one of its key aims had to be to exert what influence it could to ensure that individual European transformation efforts converged rather diverged.[1] One means for managing this feat was to develop concepts suited to the circumstances of NATO's European members, and so furnish a common framework for transformation.[2] In essence, ACT developed concepts that bridged the gap between what the US was doing and could do and what the European militaries would be able or wanted to do. Some European militaries, such as those of The Netherlands and Spain, will look explicitly to NATO to provide the model they follow, but other states will look at the range of possible concepts that already exist to interpret, develop, and implement their versions of these concepts to match their specific needs and circumstances.

Nor is it the case that the European militaries specifically look to the US as the model they should follow. Individual European militaries are certainly very much aware of the US model and its significance, and undoubtedly keep a weather eye on ongoing developments in the American military. Britain and France both look first to the US as the model to be emulated. In some instances, these two countries are pursuing emulative approaches to shared problems: Britain's Future Rapid Effects System (FRES) and France's Scorpion programs to develop medium-weight, networked army units mirror the US Army's Future Combat System (FCS) program in terms of concept and approach.[3] But though Britain and France may look first to the US, they do not import American concepts and programs as they are; rather, these two states impose their own interpretation and approaches to suit their particular circumstances and requirements. Some national military organizations, such as that of Britain, are explicitly engaged in military transformation to ensure that they can fight alongside the US, while others, such as the Royal Netherlands military, understand that they will conduct military operations only as part of a coalition, and that means being interoperable with NATO and its dominant member military, the US. What the six case studies suggest is that some states pay attention to the US model as an element of their transformation effort, as they desire to ensure that the outcome of their transformation effort, however distinctive, will be compatible with the US military's model of transformation. Other national militaries, such as the French military, may be influenced in their transformation efforts as a result of their government's ambitions with respect to the ESDP and European-only operations. Still other European military organizations, however, appear to see the US or NATO model—and language—as a means of conferring legitimacy, either to secure domestic support and resources or to impress the US and NATO allies that they can sit at the same military table.

The six case studies, however, also suggest that there are some potential points of access through which ACT may be able to shape the development of transformation in NATO Europe. While European states do not necessarily turn to ACT for these concepts (EBAO, NEC, and expeditionary-capable force), an interesting finding is that undertaking "transformation" is frequently seen as necessary in order to meet national commitments to NATO. As suggested above, a danger here is that states may be employing the language of EBAO, NEC, and expeditionary-capable force in a manner that appears consistent with NATO's use of those terms, even while the actual meaning of these concepts and the contents of implementation are inconsistent. Nonetheless, a desire to engage in transformation in order to meet NATO requirements creates an opportunity for ACT to work directly with national militaries to interpret and develop their transformation concepts and road maps. An important example of such a contribution is ACT's Multinational Experiments, which NATO member states are increasingly participating in to learn new lessons and to refine their own concepts. In working with member states, ACT has the opportunity to influence how a national military organization interprets and implements transformation, at least to the degree of ensuring that their individual transformation concepts are NATO compatible, even if only with respect to the most important measures required to enhance operational interoperability.

A second important observation is that European militaries' operational experiences may serve as a catalyst for transformation or may serve as validation of the main elements of "transformation." The Polish military was spurred to pursue military transformation by its operational experiences (and by working with other militaries seeking to employ transformation), while Britain and The Netherlands, for their part, have sought to employ an effects-based approach to their operations in Afghanistan and are using that experience to refine and develop the concept. What this suggests is that European states engaged in operations, if not already undertaking "transformation," may be more receptive to transformation if they perceive a need to improve their operational effectiveness. There thus may be an open door through which NATO can offer concepts such as EBAO, NEC, or expeditionary-capable force as solutions that, if not perfect for the needs of the particular European military, can be adapted to the specific requirements of the national military organization. Conversely, at least inductively, those European militaries not engaged in operations may be more resistant to change (or transformation) because they see no need to improve their operational effectiveness.

One final observation emerges from the six case studies. Quite clear is that particular national issues, such as national interests, domestic political climate and budgets, and military service interests, do exert a significant influence on transformation, but that one of the most substantial issues affecting transformation in Europe, if not the most substantial, are the limited resources that European militaries can devote to "transformation." Limited defense budgets seriously restrict European militaries' ability to "transform." Compounding the problem, however, is that the costs of sustaining a relatively high level of operational deployments entails a significant financial burden, which squeezes even further what they are able to devote to transformation. The current solution often opted for is to decrease the funds for acquisition or engage in force cuts, with obvious negative consequences for future operations. This problem looks set to worsen as a result of the impact of the financial crisis on European economies. Estimates that it may be several years before Europe's national economies start to emerge from the continent-wide recession raise the possibility that resources allocated to defense will be even more constrained, even as many European militaries continue to sustain costly operations in places such as Afghanistan, Bosnia, and Kosovo—among other contingencies. There is thus every likelihood that the resources that individual militaries are able to devote to transformation may be iteratively reduced in the coming years, slowing or curtailing their capacity to implement transformation and possibly even derailing their efforts entirely.

THEORETICAL CONSIDERATIONS

The six case studies of military transformation in Europe examined in this volume may still be, to a lesser or greater degree, works in progress, but the various cases, both individually and as a whole, offer a range of insights relevant to theory. The Introduction to this volume identified three main sets of literatures—military innovation, norm diffusion, and alliance theory—that furnish frameworks for understanding military transformation. Each of these approaches identifies a number of factors or processes that are applicable to examining and understanding change in military organizations. These sets of literatures are separate, ostensibly different approaches to explaining different processes and factors, yet, as the six comparative studies make clear, the dynamics involved in the process of military transformation are complex; hence, elements of one may cross-correlate to elements of one or more of the other literatures.

Military Innovation

In the Introduction we identified four main sets of hypotheses that can be derived from the literature on military innovation. These main hypotheses address issues and considerations which are key factors that shape military change: strategic imperative for change; civil-military relations; military culture; and bureaucratic interests.

Strategic Imperative

In the military innovation literature a key theoretical issue or question is, What is the source of military change? The main hypothesis is that military organizations will innovate when they perceive that they face new challenges from a change in the strategic/military environment or because operational experience has indicated that their battlefield effectiveness is inadequate or inappropriate. In the six cases here, three main strategic imperatives have motivated the instigation of military transformation: the changing security and strategic environment; constrained resources; and alliance commitment.

The European military organizations examined have all identified the need to become more expeditionary in capability and outlook, reflecting their recognition that the strategic environment has changed from one in which they had to be very concerned about the possibility of defending Europe's borders from external attack to a strategic environment in which they were being, or were likely to be, tasked to conduct operations outside of Europe's borders. The French military is perhaps an exemplar of this, having learned from its experience in the 1991 Gulf War that it needed to develop its capacity to conduct expeditionary operations, and do so alongside coalition or allied partners such as the US, if it was to be successful. The Polish military determined to develop its expeditionary capability much later than did the French, reaching this decision after multiple deployments abroad that exposed deficiencies in conducting expeditionary operations. The perceived requirement to become expeditionary, which to effect involves substantive organizational change, thus can be seen to be a rational response to the changes in the international security and strategic environment.

The shifting security environment was not, in some cases, the rationale, or sole rationale, for the development of an expeditionary capability, or indeed for the adoption of EBAO or NEC. For the French military, the embracing of the concept of military transformation—especially EBAO and NEC—was driven by a concern about the prospective decline in resources. The French military, having determined that it needed to reorient itself to conduct expeditionary operations with allies more effectively, moved further to embrace transformation, more broadly conceived as a means to make more efficient use of a seriously declining budget. The German military, for its part, recognized that not only was reorienting itself to conduct operations overseas a pragmatic strategic step to take but also that this might serve as a justification for minimizing the scope of the German government's ongoing cutting of defense resources. For both the French and German militaries, the decision to pursue military transformation was perceived as a means of sustaining military effectiveness in the face of budget cuts, or as a way to minimize as much as feasible the impact of expected reductions in the allocation of resources on their military effectiveness.

Another source of military transformation in Europe has been concern about alliance relations with NATO. The Polish military's operations abroad revealed real deficiencies in its combat effectiveness compared with the effectiveness of its NATO allies, and this in turn raised concerns that these weaknesses might have a serious adverse impact on Poland's position in the Atlantic Alliance. The source for undertaking transformation, then, was not solely the changing strategic threat environment; rather critical for Poland was ensuring that it sustained its positive relations with, and voice within, NATO. A similar concern—that is, being able to operate with other NATO forces as a source for change—is implied in the case of Spain. The Spanish military was motivated to undertake change in part by its recognition that there was an inexorable quality to military transformation. A central emphasis in the Spanish military's adoption of NEC was the recognition that this was the direction that NATO was going, and so it should move in the same direction. In short, alliance considerations were an important source of military transformation for both Poland and Spain.

Civil-Military Relations

The second main hypothesis in the literature on military innovation centers on civil-military relations. One school of thought holds that a military organization will tend to be resistant to change because of its stake in current military practices, equipment, and structures, and hence that civilian intervention is required to force a military organization to change. The other school of thought argues that military organizations will and do innovate by choice, that a visionary military leader with internal stature and legitimacy can lead a successful campaign to effect change within the military organization. In the six case studies of transformation examined in this volume, there are examples that offer support to both hypotheses.

The cases of the UK and The Netherlands fit with the first hypothesis, that military change is led or driven by the civilian leadership. The requirement for the British military to undertake change was first signaled when the newly elected New Labour government laid out a substantial rethink of the future role and shape of the British armed forces in its Strategic Defence Review (SDR) in 1997–98. In The Netherlands, as de Wijk and Osinga argue, it was primarily civilian Ministry of Defence policymakers who drove the process of transformation. In both cases, the core change enforced was to alter the mission of the military organization from its Cold War emphasis on defending Europe to conducting expeditionary operations in support of the government's foreign policy goals. As well as setting forth a new mission, both governments further framed the character, or content, of the changes to be made through the imposition of reductions in force strengths and cuts in resources. In establishing a new mission and circumscribing resources, the British and Dutch civil authorities furnished a clear impetus for their respective military organizations to innovate new approaches.

The case studies of France, Spain, and Poland provide evidence that supports the hypothesis that military organizations will choose to innovate without civilian intervention. In each case it was the military that took the initiative to argue the need to transform. The French military recognized that it needed to innovate but faced a political lobby, which included the French civilian leadership, that was wedded to an older defense paradigm that needed to be convinced of the need for change. In the cases of Spain and Poland, the military took the lead in defining the need for transformation, and of the character of the change required, in a domestic political environment in which the civilian leadership lacked expertise in military affairs. The difference between these two latter cases is that the Polish government appears to be involved in at least some defense issues, as evidenced by the Polish government's engagement in the issue of an American ballistic missile defense system on its soil (and in particular the deployment of Patriot air defense systems),[4] whereas in the case of Spain the civil leadership has demonstrated only a passing interest in military affairs. These three cases thus present two different circumstances in which the military assumed the lead. In France the civil leadership was a potential obstacle to the military change that the French military perceived as necessary, while in Spain and Poland the civil leadership represented a political void in which the two respective military organizations were relatively free to define a need for change more or less as they saw fit.

Germany's transformation effort sits somewhat askew from the two contending hypotheses. As Borchert's analysis makes clear, the initial impetus for change came from the German military, but there are limits as to how far they could take the process. Germany's political culture values consensus at the political level, with the government needing to work as a team, and consensus is a general principle adhered to within the German military as well. The absence of a broad consensus in support of military transformation creates a situation in which it is difficult for policy entrepreneurs, whether civilian policymakers or senior military leaders, to drive change. Yet Germany's strategic culture makes it difficult to develop a consensus, within the government and public, on redefining the roles and missions of the German armed forces to provide a strategic rationale for military transformation. Germany's strategic culture inclines Germany toward diplomatic and economic approaches rather than military power and toward tightly integrating its armed forces into international institutions and organizations in support of Germany's position of "international solidarity." The result is that there is little public discussion of whether and how the German armed forces should change, and the German MoD and military are "most often agenda takers, not agenda setters." This example suggests that to effect such a major military reorientation as is involved in transformation may require a substantial degree of political support to be successful.

The evidence from the six cases of military transformation offer support for both of the main hypotheses. The cases of Britain and The Netherlands suggest that governments that are attentive to the achievement of the national interest in foreign policy are more likely to perceive that changes in the strategic environment require a reorientation of its military power. The cases of France, Spain, and Poland suggest that when the political elite is not attentive to defense issues and the relationship of military power to the achievement of national foreign policy aims, or the civilian policymakers lack expertise in military affairs, a political and intellectual opening is present that the leadership of the military can, and may well, fill. What these cases, and especially that of Germany, suggest is that political culture and strategic culture provide the context that can influence whether military change is led or driven by the intervention of policymakers or by a senior military leader. In sum, the form and character of national governance, often influenced by political culture or strategic culture, can condition whether it is more likely that civilian leadership may lead, indeed, may need to drive military change, or whether military change, if it is to occur, will be led by the military leadership.

Military Culture

The third main hypothesis in the literature on military innovation is that military organizational culture will have an influence on the acceptance of change, with the literature suggesting that culture often is an important factor shaping military change. Among the six case studies there is only indirect evidence to support this particular hypothesis. In the case of Spain, Marquina and Diaz observed that individual services fought to modernize their traditional capabilities in spite of established transformation guidelines. The individual services claim for modernizing core platforms and capabilities instead of pursuing programs to meet set guidelines for transforming may have stemmed from service organizational culture, but the scope of Marquina's and Diaz's study did not permit the level of detailed investigation needed to state unequivocally that the service acquisition preferences were a function of service culture. Indirect evidence of the influence of military organizational culture may also be found in the argument of de Wijk and Osinga that the Royal Netherlands Air Force and Navy were keen promoters of transformation because the changes pursued, particularly that of NEC, were relatively consistent with their current practices, whereas for the Royal Netherlands Army transformation portended disruptive change. There are, then, suggestions among the case studies that military culture may exert some influence on the readiness of individual services to accept significant change, with the result that progress may be held back or uneven across different services.

The cases of the British military and, again, the Dutch military demonstrate another way that military organizational culture can influence military change. As Farrell and Bird point out, the British military's approach to adopting the

American concept of network-centric warfare (NCW) was shaped by its organizational culture. The US concept of NCW is very much a technology-driven concept, and this US model was inconsistent with the British military culture of "mission command" and of skepticism regarding over-reliance on technology. These cultural predispositions, along with resource constraints, shaped the British military's approach to applying information technology, resulting in its opting for "network-enabled capabilities" instead of NCW, in spite of its clear aim of operating in an integrated fashion with the US military. In the case of The Netherlands, de Wijk and Osinga point out that the Dutch military understand that it will engage in military operations only as part of a coalition. This cultural view of its purpose arguably predisposes it to look to concepts being implemented by the main coalition it likely would operate with, in this case NATO, as the source of transformation concepts, as opposed to looking to a single country for ideas. What these two examples demonstrate is that military organizational culture can shape the choices made regarding the form and content of innovations, even when those innovations are borrowed from another military organization.

The case studies of military transformation in Europe present support for the significance of the role of military organizational culture in military change. The evidence from the six cases in this volume indicates that military organizational culture influences the ease of "change acceptance." The degree of readiness to accept substantial, disruptive change will vary across specific military services because of their particular organizational cultures, depending on the character of the intended change, and will also vary across national military organizations. The evidence further indicates that a military organization, whether the national armed forces as a whole or possibly individual services, will seek to modify innovations if these changes are perceived as being contrary or harmful to their organizational culture. Put another way, military organizational culture can condition the interpretation of military concepts and ideas, influencing the meaning and content of the changes undertaken. Military organizational culture, in sum, is an ever-present factor that shapes the process of change, potentially affecting the degree of progress, the choices made, and the substance of the outcome reached.

Bureaucratic Interests and Politics

The final main hypothesis stems from the large body of literature on bureaucratic politics and on weapons acquisition. This literature establishes that there are a range of vested interests within a military organization, and within the military acquisition process, that act to sustain and enhance their position, influence, and resource share, and military changes that put these interests at risk are likely to be resisted. The overarching character of the analyses in this volume does not permit digging down to elucidate in any real detail bureaucratic and political wrangling that may have blocked or slowed the introduction and process of military transformation, yet there is evidence to support elements of this broad hypothesis.

The most obvious example of the impact of vested service interests is found in the case of Spain. Possibly the most apparent illustration is the Spanish Army's program to modernize its heavy armor capability without, as Marquina and Diaz observe, any real plans relating to how this modernized heavy armor capability would be transported abroad. Although the precise role of vested interests within the Spanish Army cannot be identified, given that Spain's MoD had already established the guidelines for making the Spanish military expeditionary capable, the general perception is that the army seemingly proceeded in a business-as-usual fashion rather than working out what it needed to do to become expeditionary capable as the basis for this acquisition program.[5] The main theoretical implication is that bureaucratic interests and politics can, and will, hinder the process of transformation and claim defense resources that could be gainfully used to implement transformation with the possible result of the acquisition of capabilities that are inconsistent or even contrary to the requirements of transformation.

The British and French armies also offer examples of the influence of vested interests. Both armies definitely embraced the need to innovate if they were to become expeditionary capable and hence able to contribute positively and effectively to the achievement of their nations' foreign policy goals. The approach both adopted was to innovate a medium-weight capability that would be capable of being deployed abroad very rapidly while still being effective in high-intensity combat operations. The basic concept adopted, mirroring that of the US Army, was based on a mix of multiple common platforms tied together by information and command networks into a "system of systems." The emulation of the US approach may have been a rational response to a perceived problem, but at the same time the British and French army leadership were very much aware that the integrated "system of system" character of their medium-weight approach meant that the civil authorities could not cut parts of the system and overall program as a cost-saving measure without the entire concept unraveling. Hence, a spur for the two militaries for the particular support adopted was that it would entail a commitment of continued funding, even should defense resources be scarce. The bureaucratic desire to obtain and sustain resources cannot be said to be the single most important reason for the British and French armies approach to becoming expeditionary capable, but the desire to secure resources clearly exerted a contributing influence on the approach those two militaries adopted.

Norm Diffusion

The literature on norm diffusion emphasizes the role of external processes of military change, unlike the military innovation literature, which focuses mostly on internal processes and factors. The norm in question here is the concept of "military transformation," and at the core of the cases examined here is that of the emulation of the US military. There are four main observations, each which suggest a hypothesis.

The first observation is that a military organization will *emulate* another that has been successful in battle either in order to improve its own military effectiveness or in order to confer legitimacy upon itself. The evidence from the six case studies indicates that both military effectiveness and legitimacy are motives that are present in European efforts.

The British and Netherlands military organizations are examples of the emulation of another military organization, because they perceived the concept of military transformation as effectively addressing strategic challenges they confront. These two military organizations recognized the requirement to be expeditionary capable set by their civilian leadership and perceived the concepts of NEC and EBAO as enhancing their military effectiveness in conducting operations overseas. Moreover, the pursuit of military transformation supports the aim of the British armed forces to be able to fight with their American ally, while it contributes to the goal of the Dutch armed forces of being able to operate in a coalition, with NATO being the main military coalition of interest. For the Polish military, the embracing of military transformation also can be seen as a rational response, for a key motive for it is the goal of ensuring that it is not perceived as a military free rider in the Atlantic Alliance. Thus, for these three military organizations the adoption of the norm of military transformation is a rational response to perceived strategic needs.

The French military, on the other hand, offers a case in which the adoption of military transformation is also a means of conferring legitimacy. The

French armed forces did undertake transformation as a rational response to the changes in the international strategic environment, but at the same time it was also motivated in part to adopt transformation, and in particular the language of transformation, as it confers on it the legitimacy of being a modern and effective military with the right to sit at the same table as the US and other allies that are implementing military transformation. The Spanish military also appears to be following the lead of NATO in military transformation in no small part for reasons of legitimacy. The Spanish armed forces started down the road of military transformation at least partly in the belief that that transformation is the wave of the future, and so they too need to "transform." Indeed, not to stretch the point too far, the rationale that "everyone else is doing it" as a motive to embark on military transformation probably applies to a lesser or greater degree across NATO European members, not least as it conveys to the US and NATO that they are endeavoring to close the transatlantic military capability and conceptual gap and to meet Alliance commitments.

There are two main theoretical points. First, in the Introduction it was suggested that the motivation for small states to pursue "military transformation" would be, given the serious constraints on their capabilities and resources, to confer legitimacy, and hence by implication that larger states, having more substantive capabilities and resources, would do so for rational strategic reasons. This aspect of the hypothesis, however, does not hold up across the six case studies. France is a large power, with reasonably substantial resources and capability, and strategic reasons of national interest for developing enhanced capabilities, yet an important motivation for the French military to emulate the US and NATO has been to confer legitimacy. On the other hand, the case of The Netherlands, though it must be considered more a medium power than a small state, does suggest that small states may undertake military transformation for rational strategic reasons to the degree to which they are able to do so. What these exceptions suggest is that the context in which emulation occurs will be an important consideration. Second, whether a state or national military organization may be motivated to adopt the norm of military transformation either for strategic or legitimacy reasons is not necessarily a strict either/or question. It may be, rather, that both rational and legitimacy reasons may impel a state to emulate the norm of military transformation, though it can be expected that one of the two motivations will probably provide a more compelling impetus and hence more powerful explanation.

The second aspect of the norm diffusion literature is the issue of *norm strength*—that is, how well defined is the norm, how long established, and how widely accepted. On the one hand, that most of the European NATO members, along with a number of European non-NATO states, are engaged in military transformation indicates that it is widely accepted. On the other hand, it is still too early to determine whether military transformation will be sufficiently sustained in the years to come to identify it as long established. The evidence from the six case studies that there are differences in the interpretation and meaning of the core concepts—expeditionary-capable force, NEC, and EBAO—suggests that military transformation is not a well-defined norm. The established character and definition of military transformation are currently further being challenged as the American and European militaries are undertaking substantive changes as a result of their recent operational experience. As one example, General James Mattis (USMC), commander of US Joint Forces Command (and of NATO Allied Command Transformation as well), effectively jettisoned in August 2008 the US concept of Effects Based Operations, the original source of EBAO, as an American operational planning approach.[6] Similarly, the US Pentagon under Secretary of Defense Robert Gates is currently working to reorient the American military away from its past focus on hi-tech conventional warfare toward being more irregular warfare capable,[7] while General Sir David Richards, the future British chief of the General Staff, has called for the UK military to adopt a similar approach.[8] Such moves do not negate necessarily the idea of military transformation, but they do raise questions about the future persistence of military transformation and about the precise definition of transformation. In short, European military organizations may accept and be engaged in the process of military transformation, but the persistence of the idea is open to question while the content or interpretation of transformation appears to be malleable and fungible. At present, then, military transformation cannot be considered to be a strong norm.

The third consideration in the norm diffusion literature relates to the *transmission structure* for new norms, with the emphasis being on the role of international institutions in promoting, transmitting, and sustaining norms. The evidence from the six cases makes clear that the key institution serving as the transmission structure of military transformation is NATO, and in particular ACT, with there being no evidence that the European Security and Defence Policy serves as a transmission pathway. Several of the states examined do look explicitly to NATO to define the requirement for military transformation and to define the encompassed main concepts of NEC and EBAO. The terminology used is certainly that propagated by NATO rather than the US, and in some cases the undertaking of transformation is perceived as a requirement for meeting national commitments to NATO.

Bilateral military relations are also important transmission pathways for military transformation. The British armed forces, as one example, have a very close bilateral relation with the US military, and these connections have been an important pathway for the transmission of ideas from the US to Britain, albeit with the British armed forces tailoring the borrowed ideas to fit their specific circumstances. What the six case studies also reveal is that while many states may look to NATO or the US for ideas, they do not look solely to those entities. Many national militaries borrow from various sources, including other European states that are engaged in implementing military transformation, adopting those ideas which best suit their particular needs and circumstance.

In general, the main transmission pathway of military transformation in Europe appears to be NATO, with its interpretation of the norm appearing to be the most dominant, but there are also other bilateral and possibly multilateral transmission pathways present as well. The implication of multiple pathways is that there are several different interpretive lenses through which the norm of military transformation is passed, with emulating states therefore having access to potentially different interpretations of the meaning and content of military transformation. The existence of different interpretations of military transformation will contribute to drift in the meaning and content of the norm, depending on which pathway the emulating state most prizes, and this will contribute to divergence in the interpretation and meaning of the norm across European states.

The final consideration in norm diffusion is the *internalization* of the norm. In the Introduction we posited that a range of domestic or national factors, such as strategic culture, military organizational culture, political imperatives, economic or resource considerations, and bureaucratic interests, would influence norm internalization. In the six case studies there is no evidence that any of these factors have resulted in the outright rejection of military transformation. It is, however, conceivable that other European states not examined may be engaged in military modernization under the cover of the rubric of transformation—that is, upgrading their capability to do better what they have always done in the same way, as opposed to changing the way they operate or adopting new approaches. Cases in which the state is engaged in military modernization under the guise of transformation would have to be interpreted as norm failure, a form of norm rejection. At present the implementation of the Spanish armed forces transformation roadmap is stalled as a result of a lack of resources, and the persistence of this problem stemming from the ongoing economic crisis may result in Spain's being reduced to having modernized rather than transformed its military. The implication is that military modernization may result as a consequence of a failure to internalize completely the norm of transformation. Underlying the impact of resource scarcity is the role of military organizational culture and bureaucratic interests. These two factors were influential in the initial steps Spain took toward transformation; the absence of strong political leadership in military affairs caused by the lack of civil expertise and focus is a void in which organizational culture and bureaucratic interests can flourish, particularly when resource share is very constrained.

In sum, there is clear evidence across the six case studies in this volume that internal factors resulted in the adaptation of the norm of military transformation. The very obvious example of this is that all of the European states have modified significantly the US version of military transformation because of their more constrained resources. Further, the Spanish case implies that resource scarcity may be a factor that can truncate transformation short of achieving the goals of the process. Other factors were also important in influencing the European states in reworking the original US concepts and ideas of military transformation. There are instances across the six cases in which strategic culture, military organizational culture, political imperatives, and bureaucratic interests were also important factors that affected how the emulating military organization interpreted military transformation. The significance of considering norm internalization is that the process of change is affected significantly by domestic factors that condition both the trajectory and the outcome of the process.

Alliance Theory

The literature on alliance theory focuses on internal alliance behavior and, at a first look, does not appear to be related to change in military organizations. Nonetheless, in the Introduction we set forth three hypotheses derived from this literature that might be helpful in examining military transformation in Europe.

The first hypothesis is that alliances are a means for member states to produce military efficiency, and in particular this aspect is relevant to the transatlantic military capabilities and concepts gaps, which will impact adversely on operational effectiveness. This issue is manifest in the current operations in Afghanistan, where there has been to date a lack of cohesiveness to pull together the many efforts to effect the achievement of a defined strategic goal. In part this problem is a function of the lack of unity of command stemming from the US/NATO command structure that must account for some forty different national military contingents, multiple lines of command, diverse range of national restrictions, fragmented communication and transportation systems, and a lack of common operating concepts.[9] Allied Command Transformation has been vested by the Alliance member states to develop common operating concepts and has served as the institutional promoter of a concept of military transformation adapted from the original US concept. Further, a crucial aim of The Netherlands armed forces in pursuing transformation is to ensure that they can operate effectively with their allies; and Poland, based on its operational experiences, also seeks to improve its capability to be an effective contributor to alliance operations. NATO's member states at least in principle see the alliance as an institution for developing military efficiency, with at least some members seemingly committed to achieving this goal. Nonetheless, as discussed in the first section, there appears to be emerging a military capability and a military conceptual gap within Europe stemming from uneven progress in implementing transformation and from divergences in the interpretation of transformation. In part this seeming disjunction stems from the reality that this point probably holds more in the case of smaller NATO members than it does for larger member states that have distinct national perspectives, ambitions, and policies concerning security and defense.

The second hypothesis that was suggested in the Introduction as being germane was that the behavior of a state may be influenced by the enduring problem of alliance burden sharing, with the dilemma of entrapment or abandonment being of particular significance. Poland's desire to enhance its military effectiveness and hence alliance contribution lest it lose support and influence in NATO councils, provides strong support for the relevance of this hypothesis. As discussed earlier, concern about alliance abandonment was a key motivation for Poland's determination to undertake military transformation. Further, the general view that military transformation on the part of the European members is needed to close the transatlantic capabilities and conceptual gaps can be seen also to be consistent with a latent fear of abandonment, though in this broader sense such concerns may be at best only a subsidiary incentive for states to engage in military transformation. This hypothesis, as noted earlier, thus offers important insight into a strong strategic rationale that may motivate a state to undertake military change which is not directly addressed in the other literatures.

The final hypothesis, or rather set of hypotheses, set forth are derived from the two-level game involving domestic and alliance politics. There is no substantive evidence from the six cases of a member state's using alliance commitments to confer internal legitimacy for the national government. It was suggested that this particular hypothesis might be evident in new member states, but there is no evidence that Poland, which joined the alliance in 1999, today requires NATO membership to ensure governing legitimacy. In the German case, there is an aspect of alliance membership and commitments providing legitimacy for the German armed forces operating abroad, given Germany's strategic and political culture, which may be considered a subtle variation of this hypothesis. Similarly, The Netherland's MoD uses NATO defense planning documents, including those on transformation, to confer legitimacy for its plans and policies vis-à-vis the Dutch Parliament. Another suggested hypothesis was that domestic politics might produce pressures that run counter to the direction of alliance policy. There is an element of this in the case of Germany, where the German public's wariness of foreign operations, especially when they involve combat, conditions the ends toward which military transformation is directed. Internal politics are also relevant in the case of Spain, though there the evidence is that the domestic political environment is such that defense and military affairs are not publicly debated, and the current government lacks expertise in these issues. The key point of the observations is that domestic politics may or may not be a critical factor, but the evidence is that the domestic political context within which military change occurs needs to be accounted for.

To conclude, the six comparative cases of military transformation in Europe furnish empirical observations that extend and challenge the theoretical literature. Two general points relating to theory advancement that can be drawn from the observations are worth noting in closing. First is the significance and influence of resources in military transformation in Europe. The issue of resource access or scarcity figures as a source of military change (for example, the French case), a factor shaping the content of military change (the British case), and a factor interrupting the process of military transformation (the Spanish case). Simply put, the issue of resources looms very large in the contemporary processes of European military transformation, yet this factor is underplayed in the literature on change in military organizations.

Second is the complementary nature of the three main sets of literature. The norm in theorizing is to test comparative theories to determine which provides a better, or more "powerful," explanation. Yet in the analysis of the six case studies it is problematic to argue whether the military innovation literature or the norm diffusion literature offers the better theoretical lens, or to deny that alliance politics, while certainly the least compelling approach because of its particular focus, does not offer important insights into military change in Europe. The first two literatures do offer comparative advantages, as they focus on somewhat different aspects; in specific cases, when the analysis is directed to elucidate specific elements, one may offer a better theoretical approach than the other. Yet the three literatures are mutually reinforcing, with insights from one set of literature potentially furnishing useful explanations that are not accounted for by hypotheses from the others. There is no easy way to pull these different sets of literatures together into one encompassing theory as a way to develop a comprehensive understanding of change in military organizations. Looking across the six case studies, however, it is possible to classify the many hypotheses from these different literatures as fitting into the categories of either being *drivers* or *shapers*. The category of drivers encompasses those hypotheses, or factors, that are sources or motivations for military change. In the six case studies, these factors include strategic challenges, alliance commitment, and legitimacy (including resources). The issue of leadership, or change agents, can also be included in this category, for it is to state the obvious that without some form of leadership, whether civil or military, military change will not likely be initiated, and persistent leadership appears necessary to continue to impel change, otherwise the process may very well grind to a halt (as in the German case). The category of shapers comprises the many factors identified in the volume and discussed above that may and often do condition substantively the trajectory and content of military change. Such shaping factors include strategic and political culture, military organizational culture, resources, bureaucratic politics, and, overlapping with drivers, persistent leadership. The underlying point of this category is that the actual process of implementing military change is important, and shaping factors will likely exert a very strong influence on whether the military change undertaken reaches fruition and whether the product of the process succeeds in effectively addressing the aims that motivated military change in the first place.

Notes

1. Farrell and Terriff: Military Transformation in NATO

1. Andrew Cottey, Tim Edmunds, and Anthony Forster, eds., *The Challenge of Military Reform in Postcommunist Europe: Building Professional Armed Forces* (Basingstoke: Palgrave, 2002).

2. Anthony Forster, *Armed Forces and Society in Europe* (Basingstoke: Palgrave, 2006), 37–38.

3. David Gompert, Richard L. Kugler, and Martin C. Libicki, *Mind the Gap: Promoting a Transatlantic Revolution in Military Affairs* (Washington, DC: National Defense University Press, 1999).

4. The only major comparative study on European military transformation published to date is Gordon Adams and Guy Ben-Ari, *Transforming European Militaries: Coalition Operations and the Technology Gap* (Milton Park: Routledge, 2006). This book focuses on network-enablement of European militaries and does not discuss the development of effects-based thinking and expeditionary capabilities.

5. US Department of Defense, *Quadrennial Defense Review Report*, 30 September 2001, 16.

6. The Iraqi army in Kuwait, some 350,000 to 500,000 strong, was defeated in the field after 100 hours, with the loss of fewer than 300 coalition troops killed.

7. Michael Gordon and Bernard Trainor, *The Generals' War: The Inside Story of the First Gulf War* (London: Atlantic Books, 1995), 158.

8. In this regard, it is notable that the end of the Cold War had relatively little effect on the level of investment in US military research and development (R&D). While the (DoD)'s procurement budget fell from its Cold War high of $118 billion in 1985 to $56 billion in 1993, the R&D budget enjoyed a far more modest fall from $40 billion to $38.8 billion over the same period (all figures in fiscal year 1993 dollars). Theo Farrell, *Weapons*

Without a Cause: The Politics of Weapons Acquisition in the United States (Basingstoke: Macmillan, 1997) 137.

9. Chris Demchak has produced compelling quantitative evidence of global emulation of the US military model at a surface level (military doctrine and national investment in IT). Demchak argues that emulation is also occurring at a deep level, in terms of force posture, but that finding is contradicted by other qualitative work which suggests a much more patchy and selective pattern of actual military emulation. Chris C. Demchak, "Creating the Enemy: Global Diffusion of the Information Technology-Based Military Model," in Emily O. Goldman and Leslie C. Eliason, eds., *The Diffusion of Military Technology and Ideas* (Stanford, CA: Stanford University Press, 2003), 307–47; Emily O. Goldman and Thomas G. Manhken, eds., *The Information Revolution in Military Affairs in Asia* (New York: Macmillan, 2004).

10. Quoted in the *Transformation Planning Guidance* (TPG). See (DoD), "Elements of Defense Transformation," at http://www.oft.osd.mil/library/library_files/document_383_ElementsOfTransformation_LR.pdf.

11. Donald Rumsfeld, "Transforming the Military," *Foreign Affairs,* May/June 2002.

12. Donald Rumsfeld, excerpt from his speech at the National Defence University on 31 January 2002. Quoted in Edgar M. Johnson, "Workshop Introducing Innovation and Risk: Implications of Transforming the Culture of DOD," March 2004, at www.oft.osd.mil/library/library_files/document_384_D2967-FINAL.pdf.

13. (DoD) Directive 3000.05, 28 November 2005 (quotation from para. 4.1).

14. This is particularly evident, not surprisingly, for the US Army and Marine Corps. See *The US Army–Marine Corps Counterinsurgency Field Manual* (Chicago: Chicago University Press, 2007).

15. David Ucko, "Innovation or Inertia: The U.S. Military and the Learning of Counterinsurgency," *Orbis*, Spring 2008, 290–310.

16. In lieu of an updated COIN doctrine of their own, the British used a draft version of FM 3–24 in designing their campaign in Southern Afghanistan in 2007. The British have yet to produce an updated COIN doctrine, but the new draft NATO doctrine on COIN clearly has the imprint of FM 3–24. Land Operations Working Group, Doctrine Panel, *Allied Joint Publication for Counter-insurgency*, AJD 3.4.4., 12 November 2008, NATO/PfP Unclassified.

17. Alexander L. George, "Case Studies and Theory Development: The Method of Structured, Focused Comparison," in Paul Gordon Lauren, ed., *Diplomacy: New Approaches in History, Theory and Policy* (New York: Free Press, 1979), 43–68.

18. Secretary General, North Atlantic Treaty Organization, "MC Position on an Effects-Based Approach to Operations," MCM-0052-2006, 6 June 2006, NATO Unclassified.

19. Terry Terriff, "U.S. Ideas and Military Change in NATO, 1989–1994," in Theo Farrell and Terry Terriff, eds., *The Sources of Military Change* (Boulder, CO: Lynne Rienner, 2002), 91–118.

20. For a review of this literature, see Adam Grissom, "The Future of Military Innovation Studies," *Journal of Strategic Studies* 29, no. 5 (2006), 905–34; see also Farrell and Terriff, *The Sources of Military Change*.

21. Stephen M. Walt, *The Origins of Alliances* (Ithaca, NY: Cornell University Press, 1987); Kenneth N. Waltz, *Theory of International Politics* (Reading, MA: Addison-Wesley, 1979).

22. Kimberly Martin Zisk, *Engaging the Enemy: Organization Theory and Soviet Military innovation, 1955–1991* (Princeton: Princeton University Press, 1993).

23. Barry R. Posen, *The Sources of Military Doctrine: France, Britain and Germany between the World Wars* (Ithaca, NY: Cornell University Press, 1984), 47, 59, 79.

24. David French, *Raising Churchill's Army: The British Army and the War against Germany, 1939–1945* (Oxford: Oxford University Press, 2000), 80.

25. Posen, *The Sources of Military Doctrine*.

26. Stephen Peter Rosen, *Winning the Next War: Innovation and the Modern Military* (Ithaca, NY: Cornell University Press, 1991).

27. Theo Farrell, *The Norms of War: Cultural Beliefs and Modern Conflict* (Boulder, CO: Lynne Rienner, 2005); Lynn Eden, *Whole World on Fire: Organizations, Knowledge, and Nuclear Weapons Devastation* (Ithaca, NY: Cornell University Press, 2004); Isabel V. Hull, *Absolute Destruction: Military Culture and the Practices of War in Imperial Germany* (Ithaca, NY: Cornell University Press, 2005).

28. John A. Nagl, *Learning to Eat Soup with a Knife: Counterinsurgency Lessons from Malaya to Vietnam* (Chicago: University of Chicago Press, 2005), 215–16.

29. Terry Terriff, "Warriors and Innovators: Military Change and Organisational Culture in the US Marines Corps," *Defence Studies* 6, no. 2 (2006), 215–47; Terriff, "'Innovate or Die': Organisational Culture and the Origins of Manoeuvre Warfare in the United States Marine Corps," *Journal of Strategic Studies* 29, no. 3 (2006), 475–503.

30. Thomas U. Berger, *Cultures of Antimilitarism: National Security in Germany and Japan* (Washington, DC: Johns Hopkins University Press, 1998).

31. Graham T. Allison, *Essence of Decision* (Boston: Little, Brown, 1971); Morton H. Halperin, *Bureaucratic Politics and Foreign Policy* (Washington, DC: Brookings Institution, 1974).

32. Edmund Beard, *Developing the ICBM: A Study in Bureaucratic Politics* (New York: Columbia University Press, 1976).

33. Farrell, *Weapons without a Cause*, 67–121; Michael E. Brown, *Flying Blind: The Politics of the U.S. Strategic Bomber Program* (Ithaca, NY: Cornell University Press, 1992).

34. Nick Kotz, *Wild Blue Yonder: Money, Politics, and the B-1 Bomber* (Princeton: Princeton University Press, 1988); Kenneth R. Mayer, *The Political Economy of Defense Contracting* (New Haven: Yale University Press, 1991).

35. Martha Finnemore, *National Interests in International Society* (Ithaca, NY: Cor-

nell University Press, 1996); Jeffrey T. Checkel, "Norms, Institutions, and National Identity in Contemporary Europe," *International Studies Quarterly* 43 (1999), 83–114; Thomas Risse, Stephen C. Ropp, and Kathryn Sikkink, eds., *The Power of Human Rights* (Cambridge: Cambridge University Press, 1999).

36. Goldman and Eliason, *The Diffusion of Military Technology and Ideas*.

37. Alexander Wendt and Michael Barnett, "Dependent State Formation and Third World Militarization," *Review of International Studies* 19 (1993), 321–47; Dana P. Eyre and Mark C. Suchman, "Status, Norms and the Proliferation of Conventional Weapons," in Peter J. Katzenstein, ed., *The Culture of National Security* (New York: Columbia University Press, 1996); Theo Farrell, "Transnational Norms and Military Development: Constructing Ireland's Professional Army," *European Journal of International Relations* 7, no. 1 (2001), 63–102.

38. Legro, *Cooperation Under Fire: Anglo-German Restraint During World War II* (Ithaca, NY: Cornell University Press, 1995), 15–16.

39. Risse et al., *The Power of Human Rights*; Checkel, "Norms, Institutions, and National Identity"; Peter M. Haas, "Epistemic Communities and International Policy Coordination," *International Organization* 46 (1992), 1–35; Farrell, "Transnational Norms and Military Development."

40. Colin S. Gray, *Nuclear Strategy and National Style* (Lanham, MD: Hamilton Press, 1986); Alastair Iain Johnston, *Cultural Realism* (Princeton: Princeton University Press, 1995).

41. John S. Duffield, *World Power Forsaken* (Stanford, CA: Stanford University Press, 1998); Berger, *Cultures of Antimilitarism*.

42. Recent scholarship has reinforced the degree of variation in the strategic cultures of European states. Christoph O. Meyer, *The Quest for a European Strategic Culture: Changing Norms on Security and Defence in the European Union* (Basingstoke: Palgrave, 2006).

43. Farrell, "Transnational Norms and Military Development"; Farrell, "World Culture and Military Power," *Security Studies* 14, no. 3 (2005), 448–88; Amitav Acharya, "How Ideas Spread: Whose Norms Matter? Norm Localization and Institutional Change in Asian Regionalism," *International Organization* 58, no. 2 (2004), 239–75.

44. Steven R. David, "Explaining Third World Alignment," *World Politics* 43, no. 2 (1991), 233–56; Richard J. Harknett and Jeffrey A. VanDenBerg, "Alignment Theory and Interrelated Threats: Jordan and the Persian Gulf Crisis," *Security Studies* 6, no. 3 (1997), 112–53.

45. Mark L. Haas, "Ideology and Alliances: British and French External Balancing Decisions in the 1930s," *Security Studies* 12, no. 4 (2003), 34–79.

46. John S. Duffield, *Power Rules: The Evolution of NATO's Conventional Force Posture* (Stanford, CA: Stanford University Press, 1995).

47. Glenn H. Snyder, *Alliance Politics* (Ithaca, NY: Cornell University Press, 1997).

48. Robert D. Putnam. "Diplomacy and Domestic Politics: The Logic of Two-Level Games." *International Organization* 42 (Summer 1988), 427–60.

49. Michael N. Barnett and Jack S. Levy, "Domestic Sources of Alliances and Alignments: The Case of Egypt, 1962–73," *International Organization* 45, no. 3 (1991), 369–95.

50. David, "Explaining Third World Alignment."

51. Thomas Risse-Kappen, *Cooperation among Democracies* (Princeton: Princeton University Press, 1995).

52. Patricia A. Weitsman, "Intimate Enemies: The Politics of Peacetime Alliances," *Security Studies* 7, no. 1 (1997), 156–92; Haas, "Ideology and Alliances."

2. Osinga: The Rise of Military Transformation

1. Interview with Admiral Ian Forbes, *NATO Review* 2 (Summer 2003), at: http://www.nato.int/docu/review/2003/issue2/english/interview.html.

2. US Department of Defense, Office of the Secretary of Defense, Director Force Transformation, *Military Transformation, A Strategic Approach*, Fall 2003, 8.

3. Donald H. Rumsfeld, "Transforming the Military," *Foreign Affairs* 81, no. 3 (2002), 20–32. For a brief account, see also Kevin Reynolds, "Building the Future Force: Challenges to Getting Military Transformation Right," *Contemporary Security Policy* 27, no. 3 (December 2006), 435–71.

4. See, for instance, *Elements of Defence Transformation*, October 2004; *Military Transformation, A Strategic Approach*, Fall 2003; *Transformation Planning Guidance*, April 2003; and separate service-specific transformation road maps, all at www.oft.osd.mil.

5. Ronald Rourke, *Defence Transformation: Background and Oversight Issues for Congress* (Washington, DC: Congressional Research Service), 9 November 2006, 6.

6. This table and the next capture in different format the range of shifts identified in US Department of Defense, *2005 Quadrennial Defense Review* (Washington, 2006), vi–vii.

7. Rourke, *Defence Transformation*, 8–9.

8. For a recent brief account of this lineage, see, for instance, Peter Boyer, "Downfall," *New Yorker*, 20 November 2006, 56–65. For a lengthy study confirming this argument, see Frederick Kagan, *Finding the Target: The Transformation of American Military Policy* (New York: Encounter Books, 2006).

9. Thomas A. Keaney and Eliot A. Cohen, *A Revolution in Warfare: Air Power in the Persian Gulf* (Annapolis, MD: Naval Institute Press, 1995), 188. For the view that the full maturing of air power in the 1980s and 1990s is the real Revolution in Military Affairs/Military Technology Revolution today, see, for instance, Benjamin Lambeth, "The Technology Revolution in Air Warfare," *Survival* 39, no. 1 (Spring 1997), 65–83.

10. Richard P. Hallion, *Storm over Iraq: Air Power and the Gulf War* (Washington, DC: Smithsonian Institution Press, 1992), 252.

11. The following short overview of the lessons of Desert Storm is based on several sources. Besides Keaney and Cohen, *A Revolution in Warfare*; and Hallion, *Storm over Iraq*; see Michael R. Gordon and Gen. Bernard E. Trainor, *The Generals' War* (Boston: Little, Brown, 1995). For a well-researched counterpoint on the alleged decisive impact on Iraqi ground units, see Daryl G. Press, "The Myth of Air Power in the Persian Gulf War and the Future of War," *International Security* 26, no. 2 (Fall 2001), 5–44.

12. Hallion, *Storm over Iraq*, 205.

13. For a balanced account of the strategic air offensive against Iraq, see Richard Davis, "Strategic Bombardment in the Gulf War," in R. Cargill Hall, ed., *Case Studies in Strategic Bombardment* (Washington, DC: US Government Printing Office, 1998).

14. Daniel Gouré and Christopher Szara, eds., *Air and Space Power in the New Millennium* (Washington, DC: Center for Strategic and International Studies, 1997), xxii.

15. Eliot Cohen, "A Revolution in Warfare", *Foreign Affairs* 75 (March/April 1996): 44.

16. A growing point of concern thus became the high percentage of casualties caused by "friendly fire."

17. Keaney and Cohen, *A Revolution in Warfare*, 212.

18. Very good and concise discussions on the short history of the current RMA are, for instance, Chris Demchak, "Watersheds in Perception and Knowledge," in Stuart Croft and Terry Terriff, *Critical Reflections on Security and Change* (London: Routledge, 2000); and Andrew Latham, "Re-Imagining Warfare: The Revolution in Military Affairs," in Craig Snyder, ed., *Contemporary Security and Strategy* (London: Palgrave Macmillan, 1999).

19. Latham, "Re-Imagining Warfare," 224.

20. Tony Mason, "The Technology Interaction," in Stuart Peach, ed., *Perspectives on Air Power: Air Power in Its Wider Context* (The Stationary Office: London, 1998), 160.

21. Various sources list the advances of air power capabilities resulting from smart munitions, stealth, datalinks, and improved onboard and off-board sensors. This section combines conclusions of two specific studies: Gouré and Szara, eds., *Air and Space Power in the New Millennium*; and, in particular, Benjamin Lambeth, *The Transformation of American Air Power* (Ithaca, NY: Cornell University Press, 2000), esp. 301–3.

22. Zalmay M. Khalizad and John P. White, eds., *Strategic Appraisal: The Changing Role of Information in Warfare* (Santa Monica, CA: RAND, 1999); and Barry R. Schneider and Lawrence E. Grinter, eds., *Battlefield of the Future: 21st Century Warfare Issues* (Maxwell AFB, Alabama: Air University Press, 1998).

23 John Arquilla and David Ronfeldt, "Cyberwar Is Coming," *Comparative Strategy* 12, no. 2 (1993), 141–65.

24. Ibid., 143.

25. Ibid., 141.

26. Ibid., 144.

27. William S. Cohen, *Report of the Quadrennial Defense Review* (Washington, DC: US Defense Department, 1997), iv.

28. Joint Chiefs of Staff, *Joint Vision 2010* (Washington, DC: Department of Defense, US Government, 1997), 17. For a short explanation of JV2010, see Maj. Gen. Charles Link, "21st Century Armed Forces—Joint Vision 2010," *Joint Forces Quarterly*, No. 13, Autumn 1996, 69–73.

29. Demchak, "Watersheds in Perception and Knowledge," 180.

30. Admiral Bill Owens, *Lifting the Fog of War* (New York: Farrar Straus Giroux, 2000), 24.

31. See David Alberts, *Information Age Transformation: Getting to a 21st Century Military* (Washington, DC: US Department of Defense, CCRP Press, June 2002).

32. Paul Murdock, "Principles of War on the Network-Centric Battlefield: Mass and Economy of Force," *Parameters*, Spring 2002, 91–92.

33. Alberts, *Information Age Transformation*, 18.

34. See, for instance, John Warden, "The Enemy as a System," *Airpower Journal* 9 (Spring 1995), 40–55.

35. See, for instance, Daniel Byman and Matthew Waxman, "Kosovo and the Great Airpower Debate," *International Security* 24, no. 4 (Spring 2000), 1–38; Barry Posen, "The War for Kosovo: Serbia's Political-Military Strategy," *International Security* 24, no. 4 (Spring 2000), 39–84; Robert Pape, *Bombing to Win* (Ithaca, NY: Cornell University Press, 1996); Karl Mueller, "Denial, Punishment, and the Future of Air Power," *Security Studies* 7, no. 3 (Spring 1998), 182–228.

36. For studies on EBO, see, for instance, David Deptula, *Effects-Based Operations* (Arlington, Virginia: Air Force Association, 2001); Edward C. Mann III, Gary Endersby, and Thomas R. Searle, *Thinking Effects: Effects-Based Methodology for Joint Operation* (Alabama: Maxwell AFB, 2002); Paul K. Davis, *Effects Based Operations: A Grand Challenge for the Analytical Community* (Santa Monica, CA: RAND, 2001); *Joint Publication 3–60, Joint Doctrine for Targeting* (Washington, DC: US Joint Chiefs of Staff, 17 January 2002); and Joint Forces Command, *Effects-Based Operations, Concept Primer* (Norfolk, VA: Public Affairs Office, November 2003).

37. For an analysis of the merits of this model, see, for instance, Richard B. Andreas, Craig Wills, and Thomas Griffith, Jr., "Winning with Allies: The Strategic Value of the Afghan Model," *International Security* 30, no. 3 (2005/6), 124–60.

38. It is estimated that 8,000 to 12,000 Taliban fighters were killed. Numbers of Afghan civilian casualties vary from 800 to 3,500. See Michael O'Hanlon, "A Flawed Masterpiece," *Foreign Affairs*, May/June 2002, 49.

39. For a concise analysis, see, for instance, Norman Friedman, *Terrorism, Afghanistan, and America's New Way of War* (Annapolis, MD: Naval Institute Press, 2003), esp. ch. 10.

40. *Strategic Survey 2001–2002*, International Institute of Strategic Studies, London, 71.

41. In fact, after several days 90 percent of all targets were emergent instead of pre-planned.

42. C4ISR indicates command, control, communications, computers, intelligence, surveillance, and reconnaissance.

43. See statement of Gen. Tommy R. Franks, Commander in Chief, US Central Command, House Armed Services Committee, 27 February 2002, at www.house.gov/hasc/openingstatementandpressrelease/107th./02-02-27franks.htm.

44. Rumsfeld, "Transforming the Military"; see also Reynolds, "Building the Future Force," 256–58.

45. Cited in "Afghan Conflict Proves Effectiveness, Exposes Pitfalls of Smart Weapons," *Tribune Business News,* 5 September 2002. See also Anthony Cordesman, *The Lessons of Afghanistan: War Fighting, Intelligence, Force Transformation, Counterproliferation and Arms Control* (Washington, DC: Center for Strategic and International Studies, working draft, 21 February 2002).

46. Arthur Cebrowski and Thomas Barnett, "The American Way of War," in *Transformation Trends,* Office of Force Transformation, Washington, DC, 13 January 2002.

47. S. M. Hersh, "The Battle between Donald Rumsfeld and the Pentagon," *New Yorker,* 7 April 2003.

48. For detailed analyses of *Iraqi Freedom,* see, for instance, Williamson Murray and Robert Scales, *The Iraqi War: A Military History* (Cambridge, MA: Harvard University Press, 2003); Bob Woodward, *Plan of Attack* (New York: Simon and Schuster, 2004); and Anthony Cordesman, *The Iraq War: Strategy, Tactics, and Military Lessons* (Westport, CT: CSIS Praeger, 2003).

49. Kevin Woods, with Michael R. Pease, Mark E. Stout, Williamson Murray, and James G. Lacey, "Doomed Execution," in Thomas G. Mahnken and Thomas A. Keaney, eds., *War in Iraq: Planning and Execution* (London and New York: Routledge, 2007), 99–100.

50. See, for example, Max Boot, "The New American Way of War," *Foreign Affairs,* July/August 2003, at http://www.foreignaffairs.org/20030701faessay15404/max-boot/the-new-american-way-of-war.html?mode=print. But see also critical views such as those of MacGregor Knox and Williamson Murray, *The Dynamics of Military Revolution 1300–2050* (Cambridge: Cambridge University Press, 2001); Colin Gray, *Weapons for Strategic Effect: How Important Is Technology?* (Maxwell AFB, Alabama: Air University Press, January 2001); and Colin McInnes, "Spectator Sport Warfare," in Stuart Croft and Terry Terriff, eds., *Critical Reflections on Security and Change* (London: Routledge, 2000).

51. For a full analysis, see, for instance, John Peters et al., *European Contributions to Operation Allied Force: Implications for Transatlantic Cooperation* (Santa Monica, CA: RAND, 2001); or Lambeth, *The Transformation of American Air Power.*

52. In 2001 assessments put the costs of DCI at approximately €43 billion to acquire those capabilities that offer Europe the required expeditionary capabilities, which, on

the staggeringly high annual collective defense expenditure of 180 billion, amounted to no more than a manageable annual increase of about €4 billion in investments. See Kees Homan, Bert Kreemers, and Frans Osinga, *De Militaire Staat van de Europese Unie* (The Hague: Clingendael Research Paper, May 2001).

53. See, for a detailed account, Bastian Giegerich and William Wallace, "Not Such a Soft Power: The External Deployment of European Forces," *Survival* 46, no. 2, Summer 2004, 163–84.

54. Speech given by Jaap de Hoop Scheffer, NATO Secretary General, Munich, 4 February 2006, at http://www.nato.int/cps/en/natolive/opinions_22578.htm, accessed 30 November 2006.

55. Jaap de Hoop Scheffer, "A New NATO," speech given at the Norwegian Atlantic Committee, Oslo, 3 March 2006, at http://www.nato.int/cps/en/natolive/opinions_22552.htm?selectedLocale=en, accessed 30 October 2006.

56. See, for instance, *Istanbul Summit Communiqué*, 28 June 2004, at http://www.nato.int/docu/pr/2004/p04-096e.htm; and *Comprehensive Political Guidance*, Riga, 29 November 2006, at http://www.nato.int/docu/basictxt/b061129e.htm, both accessed on 30 November 2006.

57. Lord Robertson, "Facing a Dangerous World: Managing Change in Defence," speech given at Leeds University, 24 January, 2003, at http://www.nato.int/docu/speech/2003/s030124b.htm, accessed 6 September 2005.

58. Jaap de Hoop Scheffer, "NATO's Istanbul Summit: New Missions, New Means," speech given at RUSI, London, 18 June 2004, at http://www.nato.int/docu/speech/2004/s040618a.htm, accessed 6 September 2005.

59. Lord Robertson specifically put the Prague Summit decisions to transform in the context of the capability gap. See Lord Robertson, "Innovating in an Uncertain World," speech given at Montreal, 6 May, 2003, at www.nato.int/docu/speech/2003/s030506a.htm, accessed 6 September 2005.

60. See Michael Gordon and Gorden Trainor, *Cobra II: The Inside Story of the Invasion and Occupation of Iraq* (New York: Pantheon, 2006).

61. Admiral E. P. Giambastiani, "Transformation Is a Continuing Process, Not a Destination," at www.act.nato.int/transformation/transformation.html, accessed on 10 October 2006.

62. For this section, see *Understanding NATO Military Transformation* (Norfolk, VA: Allied Command Transformation, 2005), 1–8, at http://www.act.nato.int/media/5-Multimedia/Doclibrary/unmtbooketenglishversion.pdf, accessed 20 November 2006.

63. See Comprehensive Political Guidance, Riga Summit, 29 November 2006, at www.nato.int/docu/basictxt/b061129e.htm, accessed 30 November 2006. Part 3 of this document contains guidelines for capabilities development that directly refer to net-centric operations, expeditionary forces, and effects-based operations—that is, the key elements of the "New American Way of War."

64. I have made use of *Concepts for Allied Future Joint Operations,* Unclassified BiSC Document, Norfolk, Virginia, 20 February 2006. This is a draft version but one that has already passed through many iterations and is considered balanced and nearly complete.

65. "A New NATO," speech given by Jaap De Hoop Scheffer at the Norwegian Atlantic Committee, Oslo, 3 March 2006, at http://www.nato.int/cps/en/natolive/opinions_ 22552.htm?selectedLocale=en, accessed 30 October 2006.

66. For similar interpretation, see the pamphlet *Understanding NATO Military Transformation.*

3. Farrell and Bird: Innovating within Cost and Cultural Constraints

1. Foreword by the Secretary of State for Defence, the Right Honourable Geoff Hoon, MP, to *Delivering Security in a Changing World: Defence White Paper* (London: TSO, 2003), 1.

2. For contrasting perspectives, see Barry Posen, *The Sources of Military Doctrine* (Ithaca, NY: Cornell University Press, 1984); and Stephen Peter Rosen, *Winning the Next War: Innovation and the Modern Military* (Ithaca, NY: Cornell University Press, 1991).

3. Royal Navy, *Future Maritime Operational Concept,* NAVB/P(04)2, 4.

4. Royal Air Force, *Royal Air Force Strategy: Agile, Adaptable, Capable,* version 2, 2006, 5. Directorate General Doctrine and Development, British Army, *The Future Land Operational Concept* [*FLOC*], April 2004.

6. Chris C. Demchak, "Creating the Enemy: Global Diffusion of the Information Technology-Based Military," in Emily O. Goldman and Leslie C. Eliason, eds., *The Diffusion of Military Technology and Ideas* (Stanford, CA: Stanford University Press, 2003), 307–47.

7. Richard Mottram, "Options for Change," *RUSI Journal* 136, no. 1 (1991), 22–26; *Front Line First: The Defence Costs Study* (London: MOD, 1994).

8. *The Strategic Defence Review* (London: TSO, June 1998), paragraph 11.

9. Foreword to *Network Enabled Capability,* JSP 777 (London: MOD Publication, 2005).

10. The Joint HLOC contains seven core concepts: operate, command, inform, prepare, project, protect, and sustain.

11. *The UK Joint High Level Operational Concept* [*HLOC*] (London: MOD Publication, n.d.), paragraphs 209, 210, 306.

12. *Future Maritime Operational Concept,* 12. See also Royal Navy, *Naval Strategic Plan 2006,* 11.

13. Intelligence, surveillance, target acquisition, and reconnaissance.

14. Directorate of Air Staff, Royal Air Force, *Future Air and Space Operational Concept,* paragraphs 34, 48.

15. On this point, see Carl H. Builder, *The Masks of War: American Military Styles in Strategy and Analysis* (Baltimore, MD: Johns Hopkins University Press, 1989).

16. Geoffrey Till, "Adopting the Aircraft Carrier: The British, American, and Japa-

nese Case Studies," in Williamson Murray and Allan R. Millet, eds., *Military Innovation in the Interwar Period* (Cambridge: Cambridge University Press, 1996); the Battle of Britain, as an example of "one of the earliest battles to be largely decided by networked information," is given in *Network Enabled Capability*, 8.

17. British Army, *FLOC*, 5. This point is reiterated in British Army, The Army of Tomorrow: Doctrine and Future Concepts in the Land Environment, Army Code 71822 (London: Ministry of Defence, 2006), 5.

18. Secret Internet Protocol Router Network.

19. Coalition Enterprise Regional Information Exchange System.

20. Farrell interview with army staff officer SO2, DEC, MOD, Whitehall, London, 25 June 2007.

21. Paul T. Mitchell, *Network Centric Warfare: Coalition Operations in the Age of Primacy*, Adelphi Paper 385 (London: Routledge for the IISS, 2006); Nora Bensahel, "International Alliances and Military Effectiveness: Fighting Alongside Allies and Partners," in Risa A. Brooks and Elizabeth A. Stanley, eds., *Creating Military Power: The Sources of Military Effectiveness* (Stanford, CA: Stanford University Press, 2007), 198.

22. William Arkin, "Not Just a Last Resort: A Global Strike Plan, with a Nuclear Option," *Washington Post*, 15 May 2005, B01.

23. Joint Staff, *Capstone Concept for Joint Operations*, version 2.0 (Washington, DC: Department of Defence, August 2005), 20–23.

24. Arthur K. Cebrowski and John J. Garstka, "Network-Centric Warfare: Its Origin and Its Future," *US Naval Institute Proceedings*, January 1998, 28–35.

25. Heidi and Alvin Toffler, *War and Anti-War* (Boston: Little and Brown, 1993).

26. John Garstka, "Implementing Network Centric Warfare," *Transformation Trends*, Office of Force Transformation, January 2004.

27. *Defense Acquisitions: DOD Management Approach and Processes Not Well Suited to Support Development of the the Global Information Grid*, GAO 06-211 (Washington, DC: US General Accounting Office, 2006), 2.

28. *Network Enabled Capability*, 9–10, 12, 14.

29. Farrell and Terriff interview with John Garstka, Washington, DC, 27 April 2007.

30. Farrell interview with two Directorate of Equipment Capability (DEC) staff officers (Army OF5 and RAF SO1), King's College London, 22 April 2008.

31. DEC, "House of Commons Defence Committee (UK) Inquiry—Defence Equipment 2007—MOD memorandum," unclassified draft, 18 October 2007, 2.

32. Lt. Gen. Nick Houghton, Chief of Joint Operations, oral evidence before the House of Commons Defence Committee, 20 June 2006, Q52.

33. Director General Training and Support, Land Command, "Observations from Training," 2005, L.3.1.1.1.

34. Notes from 16 Air Assault Brigade collective debrief, Merville Barracks, Essex, 3 December 2008.

35. Gordon Adams and Guy Ben-Ari, *Transforming European Militaries: Coalition Operations and the Technology Gap* (London: Routledge, 2006), 42–43.

36. HLOC, paragraphs 401–2; for a concise description of Mission Command, see Joint Doctrine and Concepts Centre, *British Defence Doctrine*, JWP 0-01, 2d ed. (London: MOD Publication, 2001), 3–7; see also Brig. Nigel Jackson, "Command in the Networked Era," *RUSI Journal 7*, no. 3 (2005), 28–31.

37. Michael S. Sherry, *The Rise of American Air Power* (New Haven: Yale University Press, 1987); Theo Farrell, "Strategic Culture and American Empire," *SAIS Review* 25, no. 2 (2005), 8.

38. David French, *Raising Churchill's Army: The British Army and the War against Germany, 1919–1945* (Oxford: Oxford University Press, 2000), 80. French is the leading contemporary historian of the British Army in this period, and his account directly contradicts Elizabeth Kier's claim that "[t]anks, or 'those petrol things,' had little place in the British army officer's vision of war" in the interwar period. Kier, *Imagining War: French and British Military Doctrine between the Wars* (Princeton: Princeton University Press, 1997), 129.

39. Survey conducted by Theo Farrell and Terry Terriff. The sample size is 128 responses, with the largest groups being army officers (60 percent) and middle-rank officers (62 percent).

40. Interestingly, support for this proposition was strongest among army officers (73 percent agreed with the proposition, as opposed to 68 percent of air force officers and 64 percent of naval officers), which further undercuts the common view of the army's being the least technologically inclined of the services.

41. "Network Centric Operations (NCO) Case Study: The British Approach to Low-Intensity Operations, Part 2," Technical Report Version 1.0, US Department of Defense, n.d., 240.

42. Directorate General of Development and Doctrine , British Army, "Operations in Iraq: An Analysis from the Land Perspective," Army Code 71816, 2003, paragraph 603.

43. David Talbot, "How Technology Failed in Iraq," *Technology Review*, 107, Iss. 9 November 2004. On the bandwidth problem, see Leland Joe and Isaac Porche III, *Future Army Bandwidth Needs and Capabilities* (Santa Monica, CA: RAND, 2004).

44. Farrell interview with Lt. Gen. Nick Houghton, Chief of Joint Operations, U.K. Permanent Joint Headquarters, Norwood, 18 July 2007.

45. DCDS (EC)/07/04/10, "SRO for Delivery of Network Enabled Capability and Command and Battlespace Management (NEC and CBM)—FY 2007/08 Report," 14 April 2008, downgraded and amended copy passed to King's College, 22 April 2008, 1.

46. *Incorporating and Extending the UK Military Effects-Based Approach*, Joint Doctrine Note 7/06 (Shrivenham: Development, Concepts and Doctrine Centre, September 2006).

47. *Delivering Security in a Changing World*, 10.

48. The phrase "effects-based approach to operations" makes a fleeting appearance in JWP 0-01 *British Defence Doctrine*, released in October 2001, as do references to the requirement for the military to work in an integrated way with civilian governmental and nongovernmental agencies. However, this is not systematically developed, and the focus of JWP 0-01 remains very much on "war-fighting" and the Manoeuvrist Approach.

49. Bird interview with DCDC desk officer (Army SO1), Shrivenham, 9 July 2007.

50. See, for example, *Effects-Based Operations: Change in the Nature of War*, Airpower Education Foundation, 2001, available at www.aef.org/pub/psbook.pdf.

51. Michael Gordon and Bernard Trainor, *Cobra II: The Inside Story of the Invasion and Occupation of Iraq* (London: Atlantic Books, 2006), 3–23.

52. *British Defence Doctrine*, JWP 0-01, (London: HMSO, 1996).

53. This is clearly recognized by DCDC. Both Joint Doctrine Notes produced to date explicitly state that the Manoeuvrist Approach remains complementary to the British effects-based philosophy.

54. *The UK Military Effects-Based Approach*, Joint Doctrine Note 1/05 (Shrivenham: Joint Doctrine & Concepts Centre, September 2005).

55. *The Comprehensive Approach*, Joint Discussion Note 4/05 (Shrivenham: Joint Doctrine & Concepts Centre, January 2006). This was designated a discussion note, rather than a doctrine note, in a subtle but ultimately vain attempt to minimize resistance from other government departments to a military concept.

56. Ibid., 1-5–1-6.

57. Ibid., esp. Figure 3.1—Campaign Planning Schematic, 3–3.

58. Ibid.

59. For an overview of some of the more caustic comments relating to EBO at this time from Mattis as well as some other notable critics, such as Marine Corps Lt. Gen. (Ret) Paul Van Riper and Army Maj. Gen. David Fastabend, see Elaine M. Grossman, "A Top Commander Acts to Defuse Military Angst On Combat Approach", *Inside the Pentagon* 22, no. 16 (20 April 2006).

60. Gen. (Ret) Gary Luck, *Insights on Joint Operations: The Art and Science* (Suffolk, VA: Joint Warfighting Center, US Joint Forces Command, September 2006).

61. Ibid., 7.

62. Ibid., 1. This was particularly heartening given the traditional British emphasis on "mission command" and suspicion of staff-led processes noted earlier in the chapter. The exercise Joint Venture 05 was considered particularly unsatisfactory within DCDC precisely because it was "too staff-led." Authors' interview with DCDC desk officer (Army SO1), Shrivenham, 9 July 2007.

63. Luck, *Insights on Joint Operations*, 11.

64. Bird interview with DCDC desk officer (Army SO1), Shrivenham, 9 July 2007.

65. Gen. (Ret) Gary Luck, *Joint Operations: Insights & Best Practices* (Suffolk, VA: Joint Warfighting Center, US Joint Forces Command, July 2008).

66. The memo was entitled "Assessment of Effects Based Operations," and it was reproduced in its entirety under Mattis's name as "USJFCOM Commander's Guidance for Effects-based Operations" in *Joint Force Quarterly* 51, 4 (2008), 105–8.

67. Ibid., 108.

68. The key Joint Doctrine Publication revisions are JDP 01 (Joint Operations); JDP 3-00 (Joint Operations Execution); and JDP 5-00 (Joint Operations Planning). There will also be a revised addition of the capstone British Defence Doctrine. All documents should be ratified and released in early 2009.

69. A series of Bird discussions and interviews with members of the DCDC writing team over September and October 2008.

70. Farrell and Bird participation in Joint Operation Planners Courses, October and November 2007, PJHQ Northwood.

71. Farrell interview with Air Vice Marshall Andrew Walton, Deputy Commander Joint Warfare and Capability Development, Permanent Joint Headquarters, Northwood, 18 July 2007.

72. Commander British Forces, Op HERRICK 7, "Counterinsurgency in Helmand: Task Force Operational Design," TFH/ COMD/DO 7, 1 January 2008.

73. Interview with Lt. Gen. Ridgway, Chief of Defence Intelligence (2004), in Director CBM/J6, *Future conflict: insights from interviews with senior commanders (London: MOD, 2006)*, 29.

74. Farrell telephone interview with Brig. Andrew Mackay, CO 52 Brigade and formerly CO TFH/COMBRITFOR (AFG), 23 April 2008; James Holland, "The Way Ahead in Afghanistan," *RUSI Journal* 153 (2008), 46–51.

75. "Stability Operations and Principles," 7 Arm. Bde., TELIC 7, Land Lessons Database, Defence Lessons Identified Management System, accessed 29 April 2008, at the UK Land Warfare Centre, Warminster.

76. Farrell telephone interview with Mackay.

77. *Delivering Security in a Changing World*, paragraph 4.5.

78. House of Commons Defence Committee, *Defence White Paper 2003*, Vol. 1, HC 465-1 (London: TSO, July 2004), paragraph 34.

79. This point is highlighted in Colin McInnes, "Labour's Strategic Defence Review," *International Affairs* 74, no. 4 (1998), 833–34.

80. *Strategic Defence Review*, (London: TSO, June 1998),paragraphs 6, 19.

81. *The Strategic Defence Review: A New Chapter* (London: TSO, 2002), paragraphs 9, 26.

82. Def Cmte, *Delivering Front Line Capability to the RAF*, HC 557 (London: TSO, January 2006).

83. Def Cmte, *Defence Procurement*, Vol. 1, HC 572 (London: TSO, July 2004), paragraph 87.

84. Post Operational Interview with Air Commodore G. J. Howard, RAF, commander National Support Command (Afghanistan), Operational HERRICK 6–7 (July 2007–January 2008), 1 September 2008, Warminster.

85. National Audit Office, *Hercules C-130 Tactical Fixed Wing Airlift Capability*, HC 627 (London: TSO, June 2008).

86. *Strategic Defence Review*: A New Chapter paragraph 6.

87. Def Cmte, *Defence Procurement*, HC 572, paragraphs 81–84; House of Commons Defence Committee, *Future Carrier and Joint Aircraft Programmes: Government's Response to the Committee's Second Report of Session 2005–06*, HC 926 (London: TSO, March 2006), paragraphs 13–15; Sylvia Pfeifer, Alex Barker, and Chris Tighe, "MOD Faces Difficult Choices on Spending," *Financial Times*, 12 December 2008, at http://www.ft.com/cms/s/0/dfe1423a-c784-11dd-b611-000077b07658.html.

88. Indeed, questions have been raised about the Royal Navy's ability to crew two large-deck carriers, and the extent to which the carriers will be reliant on shore-based aircraft to provide sufficient air defense and air-to-air refueling cover. Andrew M. Dorman, *Transforming to Effects-Based Operations: Lessons from the United Kingdom Experience* (Carlisle, PA: US Army War College, 2008), 33.

89. *SDR: Modern Forces for the Modern World, Factsheets*, "Joint Rapid Reaction Forces—Land Assigned Forces."

90. *Delivering Security in a Changing World: Future Capabilities* (London: MOD, July 2004), paragraph 2.11.

91. *Army of Tomorrow*, 4.

92. These are identified as deliberate intervention, power projection, peace enforcement, focused intervention, and peacekeeping. British Army, *FLOC*, 6–8.

93. Rupert Pengelley, "Future Rapid Effect System Leads British Forces' Transformation," *Jane's International Defence Review*, 1 September 2003; James Murphy, "UK MOD Alters FRES Parameters," *Jane's Defence Weekly*, 15 June 2005; Tony Skinner, "Report Lambasts UK MOD Over FRES Indecision," *Jane's Defence Weekly*, 22 February 2007.

94. Andrew Chuter, "UK Studies Effect of A400M Delay on RAF Transport," *Defense News*, 12 October 2007.

95. House of Commons Defence Committee, *The Army's Requirement for Armoured Vehicles: The FRES Programme*, HC 159 (London: TSO, February 2007).

96. Farrell telephone interview with staff officer (Army SO1), FRES Programme, 23 April 2008.

97. Sylvia Pfeife, "Delays Hit MOD Armoured Vehicle Plans," *Financial Times*, 3 November 2008, at http://www.ft.com/cms/s/0/4e1155f4-a932-11dd-a19a-000077b07658.html.

98. *Defence News*, "New Armoured Vehicles for Afghanistan," 29 October 2009, at

http://www.mod.uk/DefenceInternet/DefenceNews/EquipmentAndLogistics/NewAr-mouredVehiclesForAfghanistan.htm.

99. HLOC, paragraph 704.

100. Development, Concepts, and Doctrine Centre, "The Joint Medium Weight Capability analytical concept," 24 April 2007, para. 9.

101. Farrell and Terriff interview with Paul L. Francis, Director, Acquisition and Sourcing Management, General Accounting Office, Washington, DC, 25 April 2007.

102. Paul Cornish and Andrew Dorman, "Blair's Wars and Brown's Budgets: From Strategic Defence Review to Strategic Decay in Less than a Decade," *International Affairs* 85, no. 2 (2009), 258.

103. Ashley Seager and Patrick Wintour, "Deepest Budget Cuts since 70s to Fill '£45 bn Hole,'" *Guardian*, 14 April 2009, at http://www.guardian.co.uk/uk/2009/apr/24/budget-spending-cuts-alistair-darling.

4. Rynning: From Bottom-Up to Top-Down Transformation

1. See Charles de Gaulle, *La France et son Armée* (Paris: Plon, 1938).

2. Charles de Gaulle, *Mémoires de Guerre: l'Appel 1940–1942* (Paris: Plon, 1954), 34.

3. Martin van Creveld, *The Transformation of War* (New York: Free Press, 1991).

4. See, for instance, the 2007 doctrine of the French Army, *Gagner la Bataille, Conduire à la Paix* (Paris: Armée de Terre, 2007), a document that will reappear later in this analysis.

5. See Louis Gautier, *Mitterrand et son Armée, 1990–1995* (Paris: Grasset, 1999), 150–71; for the "cruel" citation, see 163; "cultural shock," see 154.

6. *Livre Blanc sur la Défense* (Paris ; La Documentation française 1994), 107–23.

7. Gordon Adams, Guy Ben-Ari, John Logsdon, and Ray Williamson, *Bridging the Gap: European C4ISR Capabilities and Transatlantic Interoperability*, National Defense University, 2004, at http://www.gwu.edu/~elliott/faculty/articlenotes/C4ISR%20Gap.pdf, p. 17, accessed 8 August 2007.

8. *Le Livre Blanc* (Paris: Odile Jacob, 2008). See also the White Paper Commission, at http://www.premierministre.gouv.fr/information/les_dossiers_actualites_19/livre_blanc_sur_defense_875/.

9. *Engagements Futurs des Forces Terrestres.*

10. *Forces Terrestres Futurs 2025.*

11. *La Cohérence Capacitaire des Forces Terrestres* (14 April 2004), and *Les Capacités des Forces Terrestres Futures 2025* (27 May 2004). The BCSF would be the equivalent to a transformation office. However, it is not focused on "transformation" but rather long-term planning more broadly. In 2007 the BCSF was renamed Bureau Plan.

12. *La Transformation Capacitaire des Forces Terrestres.*

13. *Allocution du général d'armée Thorette, chef d'état major de l'armée de terre devant le CHEAR*, 2 March 2004.

14. *Infovalorisation des Engagements Terrestres.*

15. Ibid., 19.

16. The MDIE ultimately became too complex and cumbersome and therefore came to an end in February 2006.

17. Interviews with DGA officials, Paris, May 2007.

18. *La Révolution des Systèmes de Forces,* Doctrine no. 1, December 2003, 37–39.

19. The eight force systems were reduced to five in 2005.

20. Interview with army staff officer, Paris, May 2007.

21. Agence France Press, "Mme Alliot-Marie Mardi à Nîmes pour un Exercice Militaire à Grande Échelle," 18 November 2004.

22. Agence France Press, "Défense: Exercice pour Expérimenter la 'Numérisation' du Champ de Bataille," 23 March 2006.

23. *'Common Vision' on the Future of the German-French Brigade,* Bonn/Paris, 20 December 2004, 10.

24. Specifically, the French Army's command system (SICF) and the French special forces' communications system (Maestro) were compatible with NATO and thus took over. They remain in place as an overarching system until a new system is ready to take over.

25. Published a few months later as "Réflexions sur le Concept d'Emploi des Forces," *Défense Nationale,* November 1975, 15–26. The statement on conventional force projection appears on 20–21.

26. Jacques Chirac, "Il est ouvert à tous," *Le Monde,* 7 May 1996; see also *Le Monde,* "Jacques Chirac veut préparer la défense de la France du XXIe siècle," 23 February 1996.

27. *La Polyvalence Operationelle.*

28. Owing much to Marine Corps Gen. Krulak's vision of war. See, for instance, Gen. Charles C. Krulak, "The Strategic Corporal: Leadership in the Three Block War," *Marines Magazine,* January 1999, at http://www.au.af.mil/au/awc/awcgate/usmc/strategic_corporal.htm.

29. Interview, Paris, May 2007.

30. See the DGA's *Projet BOA (Bulle Opérationelle Aéroterrestre),* 6 June 2002, at http://www.defense.gouv.fr/dga/content/download/43686/436357/file/la_bulle_operationnelle_aeroterrestre_boa_19_boa2.pdf.

31. *Fantassin à équipement et liaisons intégrées.*

32. Interview, Paris, November 2007.

33. Laboratoire Technico-Opérationnel.

34. Interview, Paris, May 2007.

35. Commandement de la Force d'Action Terrestre (CFAT), and Commandement de la Force Logistique Terrestre (CFLT).

36. In addition, five regional commands were created to handle other issues related to the land forces (that is, recruitment, administration, and so on).

37. *Etat-Major Interarmées* (EMIA) and *Centre Opérationnel Interarmées* (COIA).

38. Didier Bolelli, "Chroniques: État-major des Armées," *Défense Nationale*, Issue 11 (2002), 169–72.

39. Interview, Paris, May 2007.

40. Centre de Planification et de Conduite des Opérations (CPCO), and État-major de Force et d'Entraînement (EMFEIA).

41. *La synergie des effets*. This concept was approved in May 2006.

42. Ibid., Section 2.2.

43. "Interview with General Lance L. Smith, Supreme Commander Allied Transformation," *NATO Review*, Autumn 2006, at http://www.nato.int/docu/review/2006/issue3/english/interview.html.

44. James N. Mattis, "USJFCOM Commander's Guidance for Effects-based Operations," *Joint Forces Quarterly* 51, no. 4 (4th Quarter 2008), 105–8.

45. Robert G. Bell, "NATO's Transformation Scorecard," *NATO Review*, Spring 2005, at http://www.nato.int/docu/review/2005/issue1/english/art3.html.

46. Berlin Plus of 1999 concerned the EU's ability to act with the assistance of NATO; a reverse Berlin Plus would presumably concern NATO's ability to act with the assistance of the EU. The Berlin Plus agreement entered into effect only in March 2003.

47. Interview with foreign ministry official, Paris, May 2007.

48. The first two MNEs (2001 and 2003) involved the United States, the United Kingdom, Germany, Australia, and Canada.

49. Wesley Clark, *Waging Modern War* (New York: Public Affairs, 2001), 236.

50. Interview with foreign ministry official, Paris, May 2007.

51. Nicolas Sarkozy, "Our Nation Must Continue to Give High Priority to Defence," *Défense Nationale*, April 2007, 31–43. Citations at 34.

52. The current general secretary of SGDN, Francis Delon, was secretary general of the White Paper Commission. The president of this commission was Jean-Claude Mallet, former SGDN (1998–2004) and former secretary of the strategic committee that guided the defense reforms of 1996. Jean-Claude Mallet also participated in the making of the 1994 White Paper.

53. *Le Livre Blanc*, 2008, 251–53.

54. Olivier Brochet (rapporteur), *L'Impact du Concept d'Opérations Réseaux Centrées sur les Capacités de notre Futur Appareil de Défense*, Centre des Hautes Études de l'Armament, Groupe B, 41ème session 2004–2005, 13.

55. Le Figaro, "L'Alliance Atlantique vit une Période de Transformation," 16 May 2002.

56. European Defence Agency, *An Initial Long-Term Vision for European Defence Capability and Capacity Needs*, 3 October 2006, at http://www.eda.europa.eu/genericitem.aspx?area=Organisation&id=146. Citations are from paragraphs 67 and 69.

57. This is the idea launched by Sarkozy-confident Pierre Lellouche in early 2008.

See "Une Europe de la Défense à Six, Propose le Député Pierre Lellouche (UMP)," *Agence France Presse*, 31 January 2008; "Huit Propositions pour Doter l'Union d'une Défense Commune," *Le Figaro*, 31 January 2008.

58. Both are available at the website of the chief of staff: http://www.defense.gouv. fr/ema/decouverte/le_chef_d_etat_major/decrets.

59. Compare Art. 5 of the 1982 decree with Art. 6 of the 2005 decree.

60. Compare the two Art. 18.

61. Compare Art. 10 and 12 to Art. 8 and 12.

62. *Politique Générale de la 'Transformation'*. This policy framework was endorsed by CEMA in April 2005 and finally approved in the course of 2006.

63. *Centre Interarmées de Concepts de Doctrines et d'Expérimentations*.

64. Interview with CICDE official, Paris, May 2007.

65. Interview, Paris, May 2007.

66. The ambition to "control violence" rather than fight wars is apparent in reigning French doctrines. See, notably, the new doctrine of the French Army (footnote 4).

67. Christian Malis and Nicolas Constant, "Rennaissance Militaire Française," *Défense Nationale*, May 2006, 21–34. For the schism, see 29–30.

5. Borchert: Networked and Effects-Based Expeditionary Forces

I am grateful to all the experts of the Bundeswehr who answered interview requests and provided insights into the ongoing transformation process as well as access to documents. Interviews for this paper were conducted in the second half of 2007. The paper was updated to reflect major developments throughout 2008 and was completed in early February 2009.

1. Horst Köhler, "Einsatz für Freiheit und Sicherheit," *Kommandeurtagung der Bundeswehr*, Bonn, 10 Oktober 2005, 6, 9, at http://www.bundespraesident.de/Anlage/original_630701/Rede-Kommandeurtagung.pdf (accessed 29 December 2008).

2. *Konzeption der Bundeswehr* (Berlin: Bundesministerium der Verteidigung, 2004), 9.

3. *White Paper 2006 on German Security Policy and the Future of the Bundeswehr* (Berlin: Federal Ministry of Defense, 2006), 103, at http://www.bmvg.de/portal/a/bmvg/sicherheitspolitik/grundlagen/weissbuch2006 (accessed 29 December 2008).

4. Ibid., 29–30.

5. Heiko Borchert, "Vernetzte Sicherheitspolitik und die Transformation des Sicherheitssektors: Weshalb neue Sicherheitsrisiken ein verändertes Sicherheitsmanagement erfordern," in Borchert, ed., *Vernetzte Sicherheit. Leitidee der Sicherheitspolitik im 21. Jahrhundert* (Hamburg: Verlag E. S. Mittler & Sohn, 2004), 53–79.

6. For a basic introduction, see also Manfred Engelhardt, "Transforming the German Bundeswehr—The Way Ahead," in Daniel S. Hamilton, ed., *Transatlantic Transformations: Equipping NATO for the 21st Century* (Washington, DC: Center for Transatlantic Relations, 2004), 91–113.

7. Gerhard Schulz und Hans Reimer, "Transformation der Bundeswehr—Der Weg in die Zukunft," *Europäische Sicherheit* 53, no. 5 (May 2004), 31–36.

8. Interview, Berlin, 23 October 2007.

9. The German study was called "Armed Forces, Capabilities, and Technology in the 21st Century." It has never been published because it was interpreted as a rival document to the defense policy guidelines that were established around the same time. The UK study is available online at http://www.dcdc-strategictrends.org.uk/home.aspx (accessed 29 December 2007).

10. Interview, Berlin, 11 October 2007.

11. "Auslandseinsätze der Bundeswehr: Lessons Learned?" Lecture by Maj. Gen. Rainer Glatz, Deputy Commander, Operational Headquarters, Politisch-Militärische Gesellschaft, Berlin, 30 October 2007.

12. Kosten deutscher Auslandseinsätze, Antwort der Bundesregierung auf die Kleine Anfrage der Abgeordneten Wolfgang Ehrecke, Paul Schäfer, Monika Knoche, weiterer Abgeordneter und der Fraktion DIE LINKE, BT-Drs. 16/10692, 22 Oktober 2008, at http://dip21.bundestag.de/dip21/btd/16/106/1610692.pdf (accessed 29 January 2009).

13. For general overviews, see Martin Bayer, "Proving Grounds. Briefing: The German Bundeswehr—50 Years On," *Jane's Defence Weekly*, 9 November 2007, 20–28; Josef Janning and Thomas Bauer, "Into the Great Wide Open: The Transformation of the German Armed Forces after 1990," *Orbis* 51, no. 3 (Summer 2007), 529–41; Nicholas Fiorenza, "Wide Aspirations. Country Briefing: Germany," *Jane's Defence Weekly*, 8 November 2006, 27–35; Franz-Josef Meiers, "Zur Transformation der Bundeswehr," *Aus Politik und Zeitgeschichte*, 23 May 2005, 15–22; Martin Wagner, "Auf dem Weg zu einer 'normalen' Macht? Die Entsendung deutscher Streitkräfte in der Ära Schröder," in Sebastian Harnisch, Christos Katsioulis, and Marco Overhaus, eds., *Deutsche Sicherheitspolitik. Eine Bilanz der Regierung Schröder* (Baden-Baden: Nomos, 2004), 89–118.

14. Georg Nolte, "Germany: Ensuring Political Legitimacy for the Use of Military Forces by Requiring Constitutional Accountability," in Charlotte Ku and Harald K. Jacobson, eds., *Democratic Accountability and the Use of Force in International Law* (Cambridge: Cambridge University Press, 2003), 231–53.

15. *Verteidigungspolitische Richtlinien* (Berlin: Bundesministerium der Verteidigung, 2003), 3–8.

16. The official translation for *Wirksamkeit im Einsatz* is operational effectiveness. That term, however, is somewhat misleading, as the overall goal of transformation is improved operational effectiveness. Therefore effective engagement is used in this paper.

17. The current official translation is Joint Commitment Staff.

18. With the establishment of a space situation center the Air Defense Command will play a key role in the Bundeswehr's future space activities.

19. *White Paper 2006*, 108–10.

20. See http://www.streitkraeftebasis.de (accessed 29 December 2008).

21. *Konzeption der Bundeswehr,* 24.

22. Interview, Strausberg, 3 September 2007.

23. Plans foresee an increase to €5.1 billion in 2010 and €6.1 billion in 2013. See *Bundeswehrplan 2009* (Berlin: Bundesministerium der Verteidigung, 2008), 26.

24. Sebastian Schulte, "German MoD abandons CSAR plans for NH90s," *Jane's Defence Weekly,* 21 January 2009, 10.

25. Interview, Berlin, 23 October 2007. For an official assessment of current capability shortfalls, see Bisherige Auswirkungen der Transformation der Bundeswehr. Antwort der Bundesregierung auf die Kleine Anfrage der Abgeordneten Elke Hoff, Birgit Homburger, Dr. Rainer Stinner, weiterer Abgeordneter und der Fraktion der FDP, BT-Drs 16/6099, 20 July 2007, at http://dip.bundestag.de/btd/16/060/1606099.pdf (accessed 29 December 2008).

26. Telephone interview, 21 January 2009.

27. *Konzeption der Bundeswehr,* 11.

28. *Teilkonzeption Vernetzte Operationsführung in der Bundeswehr* (Berlin: Bundesministerium der Verteidigung, 2006).

29. *The Implementation of Network-Centric Warfare* (Washington, DC: Office of Force Transformation, 2005), 19–21; *Network Enabled Capability* (London: Ministry of Defence, 2005), 8.

30. Originally the milestone was set for 2010, but that date has been pushed back by two years recently.

31. *Teilkonzeption Vernetzte Operationsführung in der Bundeswehr,* 17; Interview, Strausberg, 11 September 2007.

32. Interview, Strausberg, 11 September 2007.

33. Interview, Berlin, 23 October 2007.

34. Such a road map, however, is currently available in a draft status. Interview, Bonn, 4 February 2009.

35. Recently, the Response Forces Operations Command and the Bundeswehr Center for Transformation were tasked to address this issue. Telephone interview, 26 January 2009.

36. "Die Bundeswehr auf der Suche nach der besten Vernetzung," *c't Magazin,* 2 January 2009, at http://www.heise.de/ct/hintergrund/meldung/print/121063 (accessed 3 January 2009).

37. Interviews, Strausberg, 10 September 2007, Berlin, 11, 23, and 24 October 2007.

38. Interview, Berlin, 23 October 2007.

39. Doctrine, organization, training, leadership, material, personnel, facilities, and interoperability.

40. Interview, Berlin, 23 October 2007.

41. Interview, Strausberg, 11 September 2007.

42. *Konzeptionelle Grundvorstellungen zum Gemeinsamen Rollenorientierten Einsatz-*

Lagebild (GREL) für das streitkräftegemeinsame CD&E-Vorhaben Common Umbrella 05, draft, dated 10 February 2006, 6, 7, 13.

43. Ibid., 20.

44. For more on this, see Wilfried Honekamp, ed., *Concept Development & Experimentation. Erfahrungen aus der praktischen Anwendung der Methode zur Transformation von Streitkräften* (Remscheid: Re Di Roma-Verlag, 2008), 63–86, 160–67, 231–35.

45. Interview, Strausberg, 10 September 2007.

46. Interview, Berlin, 11 October 2007.

47. Since 2007 the Bundeswehr has been working on a networked security concept paper.

48. Investigations led by the office of the public prosecutor in Munich were stopped only a few months later, as no indication was found that soldiers were disturbing the peace of the dead.

49. Interview, Strausberg, 10 September 2007.

50. Interview, Berlin, 23 October 2007; telephone interview, 26 January 2009.

51. Interview, Strausberg, 10 September 2007.

52. *Assessment of Effects Based Operations,* Memorandum for US Joint Forces Command, 14 August 2008, at http://smallwarsjournal.com/documents/usjfcomebomemo. pdf (accessed 26 January 2009), also reprinted in *Parameters* 38, no. 3 (Autumn 2008), 18–25.

53. Telephone interviews, 26 January 2009.

54. *Konzeptionelle Grundvorstellungen der Luftwaffe zum Effects-Based Approach to Operations* (Bonn: Bundesministerium der Verteidigung, 2007). See also Jörg Neureuther, "Effects-Based Operations," *Europäische Sicherheit* 56, no. 4 (April 2007), 73–76.

55. It should be noted, however, that the German Army has started to work on a concept paper of its own. Interview, Strausberg, 10 September 2007.

56. Jörg Neureuther, "Erfahrungen im Multinationalen Rahmen am Beispiel der Multinationalen Experimentserie," in Honekamp, ed., *Concept Development & Experimentation,* 134–42.

57. Political, military, economic, social, infrastructure, and information.

58. Christian Rüther, "Knowledge Development im Einsatz," *Europäische Sicherheit* 57, no. 1 (January 2008), 64–67; Sönke Marahrens, "CD&E im Einsatz," in Honekamp, ed., *Concept Development & Experimentation,* 143–59.

59. *Bi-SC Knowledge Development Concept,* unpublished document, dated 12 August 2008.

60. Interview, Berlin, 23 October 2007.

61. Interview, Berlin, 22 October 2007.

62. As the first of Germany's main political parties, the CDU/CSU parliamentary group in the German Bundestag published a national security strategy in May 2008. The paper also included a proposal for a beefed-up national security council. See "Eine Si-

cherheitsstrategie für Deutschland," 7 May 2008, 12, at http://www.cducsu.de//mediagalerie/getMedium.aspx?showportal=4&showmode=1&mid=1279 (accessed 29 December 2008).

63. *Konzeption der Bundeswehr*, 10.

64. Ralph Thiele, "Transformation und die Notwendigkeit der Systemischen Gesamtbetrachtung," in Heiko Borchert, ed., *Potentiale statt Arsenale. Sicherheitspolitische Vernetzung und die Rolle von Wirtschaft, Wissenschaft und Technologie* (Hamburg: Verlag E. S. Mittler & Sohn, 2004), 34–54, here46–50.

65. *Ministervorlage und Konzept für die deutsche Teilhabe an Concept Development and Experimentation (CD&E)-Prozessen* (Berlin: Bundesministerium der Verteidigung, 2003).

66. *Teilkonzeption Modellbildung und Simulation in der Bundeswehr* (Berlin: Bundesministerium der Verteidigung, 2006); *Teilkonzeption Konzeptentwicklung und deren experimentelle Überprüfung* (Berlin: Bundesministerium der Verteidigung, 2008).

67. With the 2008 concept paper the Bundeswehr has stopped defining a multiyear Concept Development and Experimentation work program. This decision was based on the experience of allied armed forces that a multiyear program can be detrimental to program flexibility. Telephone interview, 12 February 2009.

68. Interview, Strausberg, 10 September 2007; *Teilkonzeption Modellbildung und Simulation in der Bundeswehr*, 15.

69. Interview, Strausberg, 10 September 2007.

70. *Bundeswehrplan 2009*, 16.

71. Jörg Neureuther, "CD&E within the Bundeswehr," Presentation for the 2007 CD&E Working Group Meeting, Norfolk, 23–25 April 2007.

72. Interviews, Berlin, 11 and 23 October 2007; telephone interviews, 22 November 2007 and 26 January 2009.

73. Interview, Strausberg, 10 September 2007.

74. Telephone interviews, 26 January 2009 and 12 February 2009.

75. Stefan Klenz, "Militärische Nutzung des Weltraums aus Sicht der Luftwaffe," *Europäische Sicherheit* 56, no. 7 (July 2007), 26–31.

76. See http://www.japcc.de/ (accessed 29 December 2008).

77. Wolfgang Nolting, "Charting the Course: German Navy," *Jane's Defence Weekly Supplement*, 2 May 2007, 32.

78. *Vorläufige Konzeptionelle Grundvorstellung "Die See als Basis für Streitkräftegemeinsame Operationen"* (Berlin: Bundesministerium der Verteidigung, 2006).

79. See http://www.coecsw.org/ (accessed 29 December 2008).

80. "Leuchttürme in Grau," *Griephan Briefe*, 4 April 2005.

81. See http://www.entec.org/ (accessed 29 December 2008).

82. Erhard Drews, "Keeping German Troops Well-Trained," *Training & Simulation Journal* 8, no. 2 (April 2007), 18.

83. Heinz-Hermann Meyer von Thun et. al., "Ausbildung—ein Motor der Transformation," *Europäische Sicherheit* 55, no. 8 (August 2006), 72–76; Günter Schweppe, "Transformation in der Ausbildung," *Europäische Sicherheit* 55, no. 11 (November 2006), 56–60; Karl-Helmut Jöbgen, "Der Mensch in der Transformation," *Europäische Sicherheit* 54, no. 8 (August 2005), 46–51.

84. For more on this, see Alexander Proksa, "Vernetzte Operationsführung (NetOpFü) in der Ausbildung der Luftwaffe," in Heiko Borchert, ed., *Führungsausbildung im Zeichen der Transformation* (Wien: Landesverteidigungsakademie, 2006), 150–66.

85. Martin Aguera, "German System Would Link Training, Info Centers," *Defense News*, 17 April 2006, 18; Schweppe, "Transformation in der Ausbildung," 56–61.

86. Uwe Katzky und Olav Hansen, "Transformation und Ausbildung in der Praxis: Das Beispiel der Fernausbildung Bundeswehr," in Borchert, ed., *Führungsausbildung im Zeichen der Transformation*, 176–95.

87. Interviews, Strausberg, 10 September 2007, and Berlin, 22 October 2007.

88. *Bundeswehrplan 2008* (Berlin: Bundesministerium der Verteidigung, 2007), 60–66.

89. See, for example, Andrew Chuter, "Britain Reveals Long-expected Cuts," *Defense News*, 15 December 2008, 4.

90. Interview, Berlin, 23 October 2007. See also *Bundeswehrplan 2009*, 40.

91. *Bundeswehrplan 2008*, 64.

92. In line with this theory, the following argument could be advanced: the war on Iraq was a turning point for the red-green German government. With the outbreak of the war Germany's relations with NATO and Washington cooled down. Germany preferred to strengthen the European Security and Defense Policy. But between 2002 and 2005 it was NATO, not the European Security and Defense Policy (ESDP), that was the driving force for defense transformation. The problem with this argument is that German defense transformation took off in this period despite Berlin's poor relations with the United States.

93. John S. Duffield, *World Power Forsaken: Political Culture, International Institutions, and German Security Policy after Unification* (Stanford, CA: Stanford University Press, 1998), 23. See also Timo Noetzel and Benjamin Schreer, "All the Way? The Evolution of German Military Power," *International Affairs* 84, no. 2 (March 2008), 211–23, here 218–20.

94. Herfried Münkler, "Heroische und Postheroische Gesellschaften," *Merkur* 61, nos. 8/9 (August/September 2007), 742–52.

95. Marco Overhaus, Sebstian Harnisch, und Christos Katsioulis, "Schlussbetrachtung: Gelockerte Bindungen und Eigene Wege der Deutschen Sicherheitspolitik?" in Harnisch, Katsioulis, and Overhaus, eds., *Deutsche Sicherheitspolitik*, 253–62, here, 260.

96. Sabine Collmer, "'All Politics Is Local.' Deutsche Sicherheits- und Verteidigungspolitik im Spiegel der Öffentlichen Meinung," in Harnisch, Katsioulis, and Overhaus, eds., *Deutsche Sicherheitspolitik*, 201–26.

97. For more on this, see Ralph Thiele and Gerhard von Scharnhorst, *Zur Identität der Bundeswehr in der Transformation* (Bonn: Bernard & Graefe Verlag, 2006).

98. This also explains why German politicians prefer to use moral justifications for the use of the Bundeswehr in international operations.

99. Eric Gujer, *Schluss mit der Heuchelei. Deutschland ist eine Großmacht* (Hamburg: Edition Körber-Stiftung, 2007), 87.

100. Lockheed Martin Breakfast Lecture, Berlin, 20 September 2007.

101. See http://www.open-community.eu (accessed 29 December 2008).

6. De Wijk and Osinga: Innovating on a Shrinking Playing Field

1. See "Wereldwijd Dienstbaar" (Defence White Paper, 'On Duty, Worldwide'), *Brief aan de Tweede Kamer* (Letter to the Netherlands Parliament), September 2007, p. 1.

2. P. J. Teunissen and H. Emmens, "De Krijgsmacht in International Verband" ["The Armed Forces in International Context"], ch. 4 in E. R. Muller et al., eds., *Krijgsmacht* (Alphen aan de Rijn: Kluwer, 2004), esp. 127–30.

3. Ibid., 136.

4. Here the simple and often used definition offered by Colin Gray is adhered to, which sees strategic culture as "modes of thought and action with respect to [force], derived from perception of national historical experience, aspiration for self-charac- terization, and from state-distinctive experiences." See Colin Gray, "National Styles in Strategy: The American Example," *International Security* 6, no. 2 (Fall 1981), 21–27. There is a burgeoning literature on the idea of strategic culture. As an explanatory factor for foreign and security policy it is also a controversial concept. For recent critical discus- sions of the various schools of thought, and the methodological issues involved, see, for instance, Christopher Twomey, "Lacunae in the Study of Culture in International Security," *Contemporary Security Policy* 29, no. 2 (August 2008), 338–57; and Colin Gray, "Out of the Wilderness: Prime-time for Strategic Culture," *Comparative Strategy* 26, no. 1 (2007), 1–20.

5. For some nice recent examples, see, for instance, Minister of Defence Henk Kamp, "De Toekomst van de Krijgsmacht," in *Militaire Spectator* 4 (2003), 193–202. In this ar- ticle, based on a speech, he almost literally mentions the dominant factors shaping secu- rity political thinking in The Netherlands that are discussed here. Similarly, see pp. 4–10 of the 2007 White Paper, or p. 43 of a lengthy internal memo, the *MinDef Introductiebun- del* (MOD Introduction Manual) of 22 February 2007, drafted by the Policy Division to serve as an introduction to the incoming new minister of defense, van Middelkoop.

6. J. J. C. Voorhoeve, *Peace, Profits and Principles: A Study on Dutch Foreign Policy* (The Hague: Martinus Nijhof, 1979). But scholars disagree about the existence of con- stant factors in Dutch foreign policy. See Y. Kleistra, *Hollen of Stilstaan: Beleidsveran- deringen bij het Nederlandse Ministerie van Buitenlandse Zaken* (Delft: Eburon, 2002), 42–61. For a general description of the strategic culture, see also Jan van de Meulen, Axel

Rosendahl Huber, and Joseph L. Soeters, "The Netherlands' Armed Forces: An Organization Preparing for the Next Century," in Jürgen Kuhlman and Jean Callaghan, eds., *Military and Society in 21st Century Europe* (New Brunswick: Transaction, 2000).

7. John Mueller, *Retreat from Doomsday* (New York: Basic Books, 1989), 219. For the reference to Holland, see p. 19. A wealth of studies have followed Mueller's thesis, all noting the changing attitudes in Western societies concerning war and warfare, using labels such as post-heroic warfare, spectator sport warfare, or risk-transfer warfare. See, for instance, Zeev Maoz and Azar Gat, eds., *War in a Changing World* (Ann Arbor: University of Michigan Press, 2001); Martin Shaw, *The New Western Way of War* (Cambridge: Polity Press, 2007).

8. See, for instance, Jan van der Meulen et al., "The Netherlands' Armed Forces."

9. See, for instance, the 1993 defense white paper *De Prioriteiten Nota, Een Andere Wereld, een Andere Defensie* [*Priorities Paper: A Different World, a Different Defense*], letter to Parliament, January 1993.

10. See Janne Haaland Matlary, "When Soft Power Turns Hard: Is an EU Strategic Culture Possible?" *Security Dialogue* 37, no. 1, (2006), 105–21; and Adrian Hyde-Price, "European Security, Strategic Culture, and the Use of Force," *European Security* 13, no. 4, (2004), esp. 340–41.

11. For a good discussion of these trends, see Christopher Coker, *Humane Warfare* (London: Routledge, 2001), esp. chs. 1, 4, 5.

12. See, for instance, Jan van der Meulen, "The Netherlands: The Final Professionalism of the Military," in Charles Moskos, John Allen Williams, and David R. Segal, *The Postmodern Military, Armed Forces after the Cold War* (Oxford: Oxford University Press, 2000).

13. See, for example, Jaap Bruijn and Cees Wels, eds., *Met Man en Macht, De Militaire Geschiedenis van Nederland, 1550–2000* [*The Military History of The Netherlands, 1550–2000*] (Amsterdam: Balans, 2003), 413–16; Christ Klep and Richard van Gils, *Van Korea tot Kosovo, De Nederlandse Militaire Deelname aan Vredesoperaties sinds 1945* [*From Korea to Kosovo: The Dutch Military Participation in Peace Operations since 1945*] (Den Haag: Sdu, 1999), 146. The PvdA in particular has been prolific in producing its own defense white papers advocating disinvestments in traditional high-intensity combat capabilities. See, for instance, its 1998 report *Investeren in een Nieuwe Krijgsmacht* [*Investing in a New Defense Force*].

14. For a very detailed description of the MOD planning process, see, for instance, I. M. de Jong and F. Foreman, "Planning binnen de Krijgsmacht" ["Planning within the Armed Forces"], in Muller et al., eds., *Krijgsmacht*, 377–96.

15. For a detailed account of strategy formation at the Ministry of Defence, and defense transformation during the 1990s, see Willeke van Brouwershaven, *Turbulentie en Strategisch Vermogen: Strategievorming bij het Ministerie van Defensie* [*Turbulence and Strategic Aptitude: Strategy Formation at the Ministry of Defence*] (Eburon: Delft, 1999).

16. *Defensienota Herstructurering en Verkleining; De Nederlands Krijgsmacht in een Veranderende Wereld* (*Defense White Paper "Restructuring and Reduction: The Netherlands Armed Forces in a Changing World"*], Brief aan de Tweede Kamer (letter to Parliament), Den Haag, 1991.

17. See www.nato.int/docu/handbook/2001 for a description of this concept.

18. "Prioriteitennota Een Andere Wereld, Een Andere Defensie" [Defense Priorities White Paper: "A Different World, a Different Defense"], *Brief aan de Tweede Kamer* (letter to Parliament), Den Haag, January 1993.

19. J. Hoffenaar, "De Nederlandse Krijgsmacht in Historisch Perspectief" ["The Netherlands Armed Forces in Historical Perspective"], in Muller et al., eds., *Krijgsmacht,* 50.

20. One of the authors of this chapter, Rob de Wijk, was closely involved in this process. For reference to this, see Bruijn and Wels, *Met Man en Macht,* 417; and Klep and van Gils, *Van Korea tot Kosovo,* 108.

21. For a detailed study of this process, see, for instance, van der Meulen, "The Netherlands."

22. See the report *Naar een Dienstplicht Nieuwe Stijl* [*Toward a New Style Conscription*], Commissie Dienstplicht (Committee on Conscription), The Hague, 1992.

23. *Defensiekrant,* nr. 2, 12-1-1993.

24. *Novemberbrief,* letter to Parliament by Mr. Voorhoeve, defense minister, 1994.

25. Operation *Deny Flight* was the NATO operation to enforce the UN no-fly zone in Bosnia, which began in 1993. Operation *Deliberate Force* was the NATO air campaign during the Bosnia campaign, and Operation *Allied Force* was the NATO air campaign during the Kosovo campaign. IFOR (Implementation Force) was the NATO-led multinational peace support force in Bosnia from 1994 to 1996, and was succeeded by SFOR (Stabilisation Force). KFOR (Kosovo Implementation Force) is the NATO-led peace support force in Kosovo.

26. *Defensie Nota 2000,* letter to Parliament by Mr. De Grave, defense minister, The Hague, 29 November 1999.

27. See "Speech by Mr. De Grave, defense minister, on the *Hoofdlijnen Notitie,*" 25 January 1999, at www.mindef.nl/speeches (accessed 21 May 2007).

28. See, for instance, his speeches on 18 May 1999 on the future of the EU defense policy and at the US National Defense University, Washington, on 26 September 2000 on transatlantic security cooperation, the one at the Royal United Services Institute on 28 November 2001 on "ESDP as a Framework for Defence Cooperation," and his letter to Parliament of 8 November 2002 on ESDP and the PCC, all at www.mindef.nl (accessed 18 October 2007).

29. See his speech at the Defence Planning Symposium in Oberammergau of 15 January 2001, published in the *Militaire Spectator* 3 (2001), 113–16. The report published by the AIV (Advisory Council for International Issues) in 2003 on *European Military*

Cooperation, Opportunities and Limitations advised the government to entertain modest expectations as far as savings and real capability improvements that could be achieved in this manner.

30. This is evident in documents detailing requirements concerning, for instance, an additional DC-10 transport aircraft (letter to Parliament, 8 March 2003), precision weapons for the F-16 (letter to Parliament, 25 February 2003), the Patriot PAC-3 update (letter to Parliament, 13 February 2004), and the Medium Altitude Long Endurance UAV (letter to Parliament, 20 September 2006).

31. MOD letter to Parliament, 30 June 2003. This returns in the MoD letter to Parliament (titled the *Prinsjesdag Brief*) of 16 September 2006.

32. Adapted from J. A. M. Oonincx and R. A. W. Thuis, in Muller et al., eds., *Krijgsmacht*, 542.

33. *Rijksbegroting Defence 2009* (2009 Defence Budget), The Hague, 16 September 2008, 18.

34. *In Dienst van Nederland, In Dienst van de Wereld* [*In the Service of The Netherlands, In the Service of the World*], Defence White Paper of the PvdA (Labour Party), November 2007.

35. For the detrimental effects of this tradition on effective defense planning, see, for instance, the good historical overview by F. J. J. Princen and M. H. Wijnen, "Defensieplanning," *Militaire Spectator* 12 (2004), 616–28.

36. Klep and van Gils, *Van Korea tot Kosovo*, 159.

37. See, for instance, a recent article coauthored by the commander of the GE/NL Army Corps, J. A. van Diepenbrugge, with Maj. Gen. R. Baumgärtel, and Major H. J. P. Corstens, "1 (German/Netherlands) Corps, Bi-national Driver for Multinational Change," *Militaire Spectator* 12 (2005), 537–42; and Major R. F. M. Schröder, DEU/NLD Common Army Vision on Future Co-operation," *Militaire Spectator* 12 (2005), 521–30.

38. See, for instance, the publication of the Ministry of Economic Affairs, *Eindrapport Sectoranalyse Defensie Gerelateerde Industrie* [*Final Report Sector Analysis of the Defense Related Industry*], Leiden, 26 March 2004.

39. For one account of this process, see Erwin van Loo, *Crossing the Border: The Royal Netherlands Air Force after the Fall of the Berlin Wall* (Den Haag: Sdu, 2003), 41–42.

40. Adapted from Teunissen and Emmens, "De Krijgsmacht in International Verband," 126–27.

41. Princen and Wijnen, "Defensieplanning," 621.

42. For a detailed discussion of the lack of jointness, see P. W. C. M. Cobelens and K. A. Gijsbers, "Gezamenlijk en Gecombineerd Optreden van de Krijgsmacht," in Muller et al., eds., *Krijgsmacht*, 663–78.

43. The air force was most advanced in that respect, while the navy had no process in place at all. For a detailed discussion, see R. H. Sandee and P. W. W. Wijninga, "Lessen uit Recente Operaties" ["Lessons from Recent Operations"], in Muller et al.,

eds., *Krijgsmacht,* 715–40; and the report of the Algemene Rekenkamer (Government Audit Organization), *Leren van Vredesoperaties* [*Learning from Peacekeeping Operations*] (The Hague: Sdu, 1996).

44. Bruijn and Wels, *Met Man en Macht,* 421; and Klep and van Gils, *Van Korea tot Kosovo,* 112.

45. Bruijn and Wels, *Met Man en Macht,* 418.

46. See Klep and van Gils, *Van Korea tot Kosovo,* 116–20, for a detailed description of these developments.

47. For its foundation, see W. A. M. van der Pol, "1 (GE/NL) Corps: het ontstaan" ["The Creation"), *Militaire Spectator* 9 (1995), 397–401.

48. Klep and van Gils, *Van Korea tot Kosovo,* 141–44. A wealth of studies was produced after the Srebrenica disaster. For an early assessment of the consequences of peacekeeping operations in the Balkans, see, for instance, J. L. Soeters and J. H. Rovers, eds., *The Bosnian Experience* (Breda: Netherlands Annual Review of Military Studies, 1997).

49. P. H. de Vries and K. A. Gijsbers, "Doctrine on the Move," *Militaire Spectator* 9 (1995), 411–20.

50. Interview with Maj-Gen. Koen Gijsbers (former commandant of 11 Air Mobile Brigade), 29 December 2007.

51. See Gen. Charles C. Krulak, "The Strategic Corporal: Leadership in the Three Block War," *Marines Magazine,* January 1999, 18-22.

52. In Klep and van Gils, *Van Korea tot Kosovo,* 155.

53. This section is informed by a good survey of recent army developments by army officers O. P. van Wiggen, C. A. M. van Eijl, and P. Nieuwenhuis, "Het Transformatieproces van de KL in Perspectief" ["The Transformation of the NL Army in Perspective"], *Militaire Spectator* 10 (2004), 481–92.

54. For a comprehensive study of changes taking pace from 1990 to 2000 within the Royal Netherlands Air Force, see Erwin de Loo, *Crossing the Border.*

55. To illustrate, air force pilots receive their basic helicopter and jet training on US-based facilities. In addition, since the late 1980s, F-16s participate in Red Flag exercises, and fighter weapons instructors are trained on a course modeled after the USAF. Dutch Patriot units participate in US-based *Roving Sands* exercises and together with US and German units organize the unique biannual Optic Windmill TBM exercise.

56. See his speech of 18 November 1999 on "The State of Our Defence," and his speech at the Air Power Conference of 7 May 2001, at www.mindef.nl/actueel/tospraken, 2001 (accessed 18 October 2007).

57. One of the authors of this chapter, Frans Osinga, was involved in various discussions revolving around the JSF project ongoing at the air staff at the time.

58. Interview with the Navy Board, 15 October 1996.

59. See his op-eds in two leading daily journals, the *NRC* of 7 May 2003, and the *Volkskrant* of 23 November 2004.

60. Clingendael Centre for Strategic Studies report *Visie op de Toekomstige Opper-
vlakte Vloot van de Koninklijke Marine* [*Vision for the Future Surface Fleet of the Navy*],
The Hague, April 2004.

61. *Luchtverdedigings—en Commandofregat.*

62. As the MOD director of Policy Evaluation dryly remarked, a proper assessment
along the stated criteria would actually have resulted in a rather different recommendation; H. P. M. Kreemers, "Beleid, Uitvoering en Evaluatie" ["Policy, Execution and Evaluation"], *Militaire Spectator* 9 (2006), 405.

63. See Klep and van Gils, *Van Korea tot Kosovo*, as well as Bruijn and Wels, *Met Man
en Macht*, for an elaborate description of these problems. The deployment in the Scre-
brenica debacle, the UNMEE operation in Ethiopia, the deployments to Haiti (1995),
and Cambodia (UNTACT, 1993) all involved various types of incidents for which the
minister of defense had to account for himself in front of the media and the Parliament.

64. One of the authors of this chapter, Rob de Wijk, was a member of the Franssen
committee.

65. Princen and Wijnen, "Defensieplanning," 627. For this description, see also A. P.
P. M. van Baal, "Besturing van de Krijgsmacht" ["Management of the Armed Forces"],
in Muller et al., eds., *Krijgsmacht*, 353–75; T. Bijlsma and E. J. de Waard, "Innoveren in
Plaats van Imiteren," *Militaire Spectator* 12 (2004), 605–15; and BGen. A. De Ruiter and
MGen. R. Bertholee, "Bestuursvernieuwing, een Stand van Zaken" ["Defense Manage-
ment Changes: The State of Affairs"], *Militaire Spectator* 11 (2005), 449–58.

66. A series of articles, for instance, discussed the efforts ongoing within the US
Army toward digitization, after the authors had presented their findings to the NL Army
Board. See N. Le Grand, D. Bongers, and M. van Maanen, "Het Gedigitaliseerde Ge-
vechtsveld in 2010" (Parts I, II, and III), in *Militaire Spectator* 9 (2001), 453–473.

67. For instance, the 2003 White Paper includes a discussion of "military-opera-
tional changes" (p.19) that reflects these insights, including a reference to the notion
that advanced military technology allows for the achievement of specific effects with the
employment of stand-off systems.

68. See, for instance, MOD letter to Parliament, 15 March 2000, concerning the F-16
replacement, the MOD letter to Parliament, 12 November 2004, regarding the ongoing
F-16 modifications, and MOD letter to Parliament, 16 March 2005, concerning additional requirements for F-16 Laser Target Designation Equipment; see also MOD letter
to Parliament, 12 August 2004, concerning the requirement for a new F-16 Reconnaissance System, all at www.mindef.nl. (accessed 16 October 2007).

69. An early entry in this debate was a three-part series of articles by Frans Osinga
entitled "Netwerkend Ten Strijde" ["Networking into Battle"] in the *Militaire Spectator*
in 2003–4.

70. See, for instance, his speech of 1 March 2004 for the KVBK on "The Future of

Our Armed Forces," or of 10 December 2004 on "The Future of Land Operations," both at www.mindef.nl/actueel/toespraken/2004 (accessed 18 October 2007).

71. These studies are classified "for internal use" only. The authors are in possession of these studies. The flyer is available at www.mindef.nl.

72. ISIS is a generic C2 system for headquarters and other semistatic commanding elements for land-based operations. It provides a Common Operational Picture (COP) to the users in order to create shared awareness in operations. TITAAN—Theatre Independent Tactical Army and Air Force Network—is a Communications and Information Systems (CIS) infrastructure, designed for the mobile military environment. The Advanced Fire Support Information System (AFSIS) is part of the brigade fire support system. THEMIS is a system for military messaging in an operational domain that automatically distributes, registers, and archives formal (military) messages.

73. Interview with Peter Krijgsman by Ward Venrooij, in Ward Venrooij, *The Diffusion of Network Enabled Capabilities in the Dutch Army*, draft research paper, Eindhoven Technical University, December 2008, 15. Level 2 is the level that all NATO forces are assumed to possess before NEC implementation and indicates that the different components at the battlefield share some information, but there is still a strong hierarchical command structure with an emphasis on preplanned missions and little scope for dynamically responding to battlefield changes. Commands deconflict their actions in time and space. Level 3 implies that there is some joint planning and that some synchronization between commands and units is possible as a result of the introduction of networking capabilities. Actions are coordinated.

74. Ibid., 23. The journal *Carré*, for instance, published a NEC special issue in August 2008.

75. Venrooij, ibid., notes twenty-three equipment programs that refer to their connection with the MOD NEC policy and NEC principles. These include requirements for F-16 link 16, F-16 aerial reconnaissance system, AH-64 upgrade, mortar detection radars, replacement of tactical army radios, replacement of satellite communication equipment, soldier equipment modernization, combat ID equipment, TITAAN C2 system, Patriot missile replacement and PAC III upgrade, a naval TBMD-related IT system, and a new marine corps CIS system. All these can be found on www.mindef.nl.

76. Letter to Parliament, *Antwoorden op de vragen van de Vaste Kamercommissie voor Defensie over de defensiebegroting* [*Replies to Questions of the Permanent Paliamentary Defense Committee concerning the Defense Budget*], The Hague, 17 October 2008, 24–26.

77. See the 2007 White Paper: "Wereldwijd Dienstbaar," 46.

78. Letter to Parliament, *Antwoorden op de vragen van de Vaste Kamercommissie voor Defensie over de defensiebegroting*, The Hague, 17 October 2008, 25.

79. *Rijksbegroting Defensie 2009*, 19.

80. See, for instance, Frans Osinga, "Revolutie in de Lucht; Air Power in het Postmoderne Tijdperk" ["Revolution in the Air: Air Power in the Postmodern Era"], *Mili-*

taire Spectator, June 2003, 338–57. An early and lonely army-oriented article was F. J. G. Toevank, B. J. E. Smeenk, and M. J. M. Voskuilen, "Denken in Effecten maakt KL-Beleid Bestendiger" ["Effects Based Thinking Makes Army Policy More Resilient"], *Militaire Spectator* 2 (2002), 102–9.

81. MOD *NEC Study*, The Hague, 2005, 53.

82. Chief of Staff, Royal Netherlands Army, *Army Doctrine Publication II—Part C, Combat Operations against an Irregular Force*, The Hague, 2003. 573.

83. Ministerie van Buitenlandse Zaken, Ontwikkelingssamenwerking en Defensie, Notitie: *Wederopbouw na Gewapend Conflict*, March 2005. For the emerging consensus in policy and practice in this field within The Netherlands government over the past decade, see, for instance, Kees Homan, "De Krijgsman als Ontwikkelingswerker?" *Atlantisch Perspectief*, November 2005, 240–48.

84. This process started in 2001. A formal re-evaluation of foreign policy noted that dealing with most challenges to internal and external security, such as terrorism, humanitarian disasters, migration, intrastate conflicts, as well as global issues such as poverty and climate change, required closer cooperation and a more integrated approach between ministries at the governmental level, as well as with NGOs and IOs. See also, for instance, Ministry of Foreign Affairs Press Communiqué of 16 September 2003, *Foreign Policy 2004*, www.minbuza.nl (accessed 20 January 2009); and the policy priorities listed in the annual report of the Ministry of Foreign Affairs 2006, www.rijksbegroting.minfin. nl.

85. For an inside account of the views concerning the 3D approach at the Ministry of Foreign Affairs, see an article by its director for conflict prevention, Robbert Gabriëlse, "A 3D Approach to Security and Development," *Quarterly Journal*, Summer 2007, 67–73. It confirms the idea of emergent consensus among the various ministries based on operational experience.

86. *Nederlandse Defensie Doctrine*, Den Haag, 2006, 64–65.

87. See *To EBO or Not to EBO?* MOD Study, The Hague, 31 July 2007.

88. Ministry of Foreign Affairs, letters to Parliament concerning NATO ministerial meetings on 23 April 2007 and on 7 December 2007, respectively.

89. *Rijksbegroting Defensie*, 16 September 2008, 11.

90. They also tested a NC3A-developed "toolkit" for these analyses. For these developments, see F. J. G. Toevank and R. G. W. Gouweleeuw, "Operationeel Analisten bij ISAF III" ["Operational Analysts with ISAF III"], *Militaire Spectator* 10 (2004), 473–80; M. Duistermaat et al., *Analysing Operational Effects*, TNO Report, June 2007.

91. See remarks concerning the conference by T. W. Brocades Zaalberg, "To COIN or Not to COIN," *Militaire Spectator* 177, no. 3 (2008), 120–23.

92. W. S. Rietdijk, "De Comprehensive Approach in Uruzgan," *Militaire Spectator* 177, no. 9 (2008), 472, 478–80. For other observations on PRT experiences, including the need for the 3D approach, see the special issue of *Carré*, January 2008.

93. Ministry of Defence, Defence Staff, Doctrine Branch, Joint Doctrine Bulletin 2008/01, *Provincial Reconstruction Teams, Deployment in Afghanistan*, www.yourdefence. nl (accessed 20 January 2009).

94. See Alfred van Staden, *Een Trouwe Bondgenoot; Nederland en het Atlantische Bondgenootschap* [*A Loyal Ally: The Netherlands and the Atlantic Alliance*] (Baarn: In den Toren, 1974).

95. With the enhanced capacity of the Defence Staff, a greater influence from NATO on the MOD planning process is expected. See de Jong and Foreman, "Planning binnen de Krijgsmacht," 396.

96. C. M. Megens, "Nationale Wensen en NAVO-Eisen ten Aanzien van de Nederlandse defensie, 1950–1990" ["National Wishes and NATO Demands concerning the Dutch Defense, 1950–90"], *Militaire Spectator* 9 (2001), 474–79.

7. Marquina and Díaz: The Innovation Imperative

1. At that time the Spanish Communist Party defended nonalignment for Spain, keeping the status quo and maintaining for a period the US military bases. The Spanish Socialist Party defended neutrality and nonalignment for Spain, as well as keeping the US military bases. The question was that since the agreements of 1953, the US bases in Spain had been oriented to the defense of Europe, not to the defense of Spain. The sectarian approach was one of the sad ingredients of that period. The strong and irresponsible campaign of nonintegration in NATO was used by the Socialist Party to win the elections in 1982. NATO was an extremely divisive issue in public opinion. With this campaign, the Socialist Party could change public opinion in three years' time. According to polls, a majority of 57 percent supported Spanish entry into NATO in 1978. By 1981, that majority was reduced to 17 percent. Javier Solana was the anti-NATO campaign chief in the Socialist Party. Later he was promoted by the US to the post of secretary general of NATO. See in this regard: Antonio Marquina, *Defense and Security in the Programs of Spanish Political Parties* (Madrid: INCI, 1980); Antonio Marquina, "Spanish Foreign and Defense Policy since Democratization," in Kenneth Maxwell, ed., *Spanish Foreign and Defense Policy* (Boulder: Westview Press, 1991), 19–62.

2. It speaks volumes of the lack of critical approaches by the Spanish military in a period where critics implied nonpromotion. Laudation or self-censorship was the norm.

3. Javier Rupérez, "La Integración de España en la Estructura Militar de la OTAN: La Búsqueda del Tiempo Perdido," in Antonio Marquina, ed., *España en la Nueva Estructura Militar de la OTAN* (Madrid: UNISCI, 1999), iii–xi.

4. The use of the word "transformation" by the Spanish authorities and many military officers is inappropriate. "Transformation" is used to mean change in organization, deployment, or institutional change.

5. The National Defence Directive 1/92 established that Spain had to plan its national defense considering three possible fields of military action: the defense of its sovereignty,

including the European and NATO dimensions; international commitments and interdependence with the surrounding countries; and finally solidarity with the UN. In this regard, the guidelines for the development of the defense policy stressed Spanish participation in peace, disarmament, and arms control, UN initiatives, and peacekeeping missions and humanitarian aid. The new National Defence Directive 1/96 signed by Prime Minister Aznar did not substantially change that approach.

6. *Libro Blanco de la Defensa* (Madrid: Ministerio de Defensa, 2000).

7. Many times in books and in official discourses the term "modernization" is utilized as synonymous with transformation. See, for instance, J. Ortega Martín, *La Transformación de los Ejércitos Españoles* (UNED: Instituto Universitario Gutierrez Mellado, 2008); Eduardo Serra et al., *Panorama Estratégico 2006/2007* (Ministerio de Defensa Madrid, 2007).

8. *Directiva de Defensa Nacional 1/2000* (Madrid: Ministerio de Defensa, 2000).

9. *Revisión Estratégica de la Defensa* (Madrid: Ministerio de Defensa, 2002).

10. *Régimen de Personal de las Fuerzas Armadas. Law 17/1999.* The aim was to achieve "more operational, flexible, more reduced and better equipped" armed forces.

11. Juan José López Díaz, "The All-Volunteer Spanish Armed Forces," in Curtis Gilroy and Cindy Williams, eds., *Service to the Country: Personnel Policy and the Transformation of Western Militaries* (Cambridge, MA: MIT Press, 2006).

12. EMAD, *Nuevos Retos, Nuevas Respuestas. Estrategia Militar Española.* Madrid, 2003.

13. Ibid., paragraphs 37–42.

14. Jesús Rafael Argumosa, "La Estrategia Militar Española," *Revista Ejercito 767* (March 2005), 12–20.

15. Two years later the National Defence Organic Law, Article 12,3a, established that it is the responsibility of the chief of Military Staff to elaborate and define the national defense strategy.

16. In the international sphere the following points were presented:

1. Resolutely support the common security and defense policy of the European Union;

2. Participate actively in the initiatives of an enlarged and transformed NATO, in particular in the Prague Capabilities commitment and the Response Force;

3. Strengthen relations between the European Union and NATO;

4. Contribute to increasing security in the Mediterranean region;

5. Foster a solid and balanced relationship with the United States;

6. Establish closer relations as regards security, defense, and military cooperation with the countries of the Ibero-American Community of Nations; and

7. Step up defense diplomacy by promoting mutual confidence with the armed forces of countries in areas of strategic interest.

17. During the first Popular Party government, Admiral Francisco José Torrente was the director general for policy at the Ministry of Defence and Gen. Felix Sanz Roldán was the deputy director for international affairs at the Ministry of Defence during the two mandates of the party. The continuity of the initiatives already promoted was thus a fact.

18. The government has spent some money on the expansion (popularization) of the "Culture of Defence" in the last ten years. But the quality of some of the initiatives that were financed and of many publications is questionable. Politicization, very questionable nominations from a scientific point of view, lack of continuity and consistency, surveillance and control by the intelligence services, and, in general, lack of excellence are endemic problems.

19. *Orden Ministerial* [ministerial order] *37/2005*.

20. *Orden Ministerial* [ministerial order] *1076/2005*.

21. Felix Sanz Roldan, "La Transformación de las Fuerzas Armadas Españolas," Conference at Club Siglo XXI, 6 June 2005.

22. In December 2008, the new minister of defense, Carmen Chacón, explained in the Defence Committee of the Parliament that the figure of 3,000 troops was obsolete. She mentioned a Spanish deployment capacity of 7,700 troops in six possible scenarios: two principal scenarios and four less demanding scenarios. See "Comparecencia para informar del desarrollo de las operaciones de paz en el exterior. Congreso de los Diputados, 10 de diciembre de 2008," at http://www.mde.es/actu_ministro/intervenciones/comparecencia_misiones.pdf.

23. Articles 17–19. In the case of maximum urgency, the approval of Parliament is still mandatory but ex post. Nothing is said regarding the discussion and approval of the mandate, the budget, the duration of the mission, and other operational questions. The government must periodically inform Parliament on "the development" of the operations.

24. This has been identified as one of the most controversial aspects in the Transformation Plan.

25. The new LHD ship, which will be ready in 2008, will have the capacity to transport more than 900 soldiers and around forty-six Leopard II or other vehicles at a distance of 9,000 miles. That, however, is still insufficient.

26. Alberto González Revuelta, "Sabemos de lo que Hablamos Cuando nos Referimos al Concepto de Transformación," *Revista Ejército* 786 (October 2006).

27. We are very grateful to the Spanish General Staff, the Unit for Armed Forces Transformation, and the CIS Division for their insights and discussions on the question of EBAO and NEC.

28. Guillem Colom, "The Spanish Military Transformation," Unidad de Transformación de las Fuerzas Armadas (Unit for Armed Forces Transformation), power point presentation made by Unidad de Transformación de las Fuerzas Armadas (Unit for Armed Forces Transformation).

29. Estado Mayor Conjunto, *Fuerzas Armadas Españolas. Mejorando la Eficacia Operativa.* Madrid, September 2008 at http://www.mde.es/descarga/FAS_Mejorando_eficacia_operativa.pdf.

30. Ibid.

31. Interview with "Unidad de Transformación de las FAS," September 2007.

32. *National Defence Directive 2004/1*, 8.

33. Félix Arteaga and Enrique Fojón, *El Planteamiento de la Política de Defensa y Seguridad en España* (Madrid: UNED, 2007), 199.

34. A good example is the almost exclusive humanitarian approach for dealing with the migration flows coming from sub-Saharan Africa to the Canary Islands. The security considerations fade away. It must be remembered that in the Spanish deployment, forces from the Ministry of Interior, the Ministry of Public Works, the Ministry of Labour and Social Affairs, and the Ministry of Defence participate. The question of borders and possible foreign penetration is not taken into consideration. There is no interdiction.

35. *Comisión de Defensa del Congreso de los Diputados*, no. 366, 13 October 2005; no. 661, 10 October 2006; no. 913, 11 October 2007.

36. "Comparecencia ante la Comisión de Defensa del Congreso de los Diputados para informar del proyecto de Nueva Directiva de Defensa Nacional," 25 November 2008, at http://www.mde.es/actu_ministro/intervenciones/081125_Comparecencia_Chacon_DDN-08.pdf.

37. For the first time, all the departments of the government making contributions to the Spanish defense have participated in the elaboration of the new directive. The former directives approved were the *domain reservé* of the Ministry of Defence.

38. "Discurso de la Ministra de Defensa en la Pascua Militar," at http://www.mde.es/actu_ministro/intervenciones/090106_Chacon_Pascua_Militar.pdf.

39. James N. Mattis, "USJFCOM Commander's Guidance for Effects-based Operations," *Parameters*, Autumn 2008, 18, at http://www.carlisle.army.mil/usawc/parameters/08autumn/mattis.pdf.

40. EBO, even being re-evaluated for the American armed forces, is a very useful concept in asymmetric conflicts, as Defense Secretary Robert Gates has remarked: "[In] coming years the main threat faced by the US military overseas will be a complex hybrid of conventional and unconventional conflicts, waged by militias, insurgent groups, other non-state actors and Third World militaries." See "Gates Speaks on Modernizing the Military," *US National Defense University*, 30 September 2008, at http://www.defenselink.mil/speeches/speech.aspx?speechid=1279.

41. That is also the opinion of the new chief of staff of the armed forces. He stated that the new military system of command and control will be slowed down, and the rest of the programs depending on the Military Staff will be reprogrammed. See "Comparecencia del General Jefe del Estado Mayor de la Defensa, JEMAD (Rodríguez Fernández)," Congreso de los Diputados, Comisión de Defensa no. 94, 7 October 2008.

42. The Spanish Navy has traditionally been considered very pro-American, to a lesser extent the Spanish Air Force, and least of all the Spanish Army.

43. The first articles dealing with EBAO and NEC are dated 2008. One possible exception is the air force. Since 2003 the workshops of Catedra Alfredo Kindelán made a contribution to this debate. However, there are no public debates in the air force magazines.

8. Osica: Transformation through Expeditionary Warfare

I would like to thank all those who shared their knowledge on military transformation in Poland and assisted me in writing this article. My special thanks go to the former and present staffers from the Transformation Department of MoND. They are, of course, not responsible for any errors I might have made in this article.

1. Chris Donnelly, *Nations, Alliances and Security: A Collection of Writings by C. N. Donnelly* (Budapest: Akademiai Kiado/ITDIS, 2004), 24.

2. *NATO Glossary of Terms and Definitions AAP-6,* 2008.

3. Peter J. Podbielski, *Transform or Modernize: Why Polish Military Transformation Matters,* 26 October 2006, CEPA, at http://www.cepa.org/publications/posts/transform-or-modernize-why-polish-military-transformation-matters.php.

4. *National Security Strategy of the Republic of Poland,* 22 July 2003, at http://merln.ndu.edu/whitepapers/Poland-2003.pdf.

5. *National Security Strategy,* Warsaw, 2007, at http://merln.ndu.edu/whitepapers/Poland-2007-eng.pdf.

6. Ibid.

7. *QDR Report,* 2001, iv, at http://www.defenselink.mil/pubs/pdfs/qdr2001.pdf.

8. Marek Ojrzanowski, "Zdolności Operacyjne—Warunkiem Skutecznych sił Zbrojnych" ["Operational Capabilities—A Precondition for an Effective Armed Forces"], *Kwartalnik Bellona* 2 (2008), 54–55.

9. Source: MoND website http://www.mon.gov.pl/pl/strona/47/LG_54_55 (accessed 30 January 2009).

10. J. Urbaniak, "Nowa Odsłona, Nowe Wyzwania," *Polska Zbrojna* 35 (2007), pp. 24-25.

11. Podbielski, *Transform or Modernize.*

12. *Jane's Defence Weekly,* 7 November 2007, 34.

13. Polish Press Agency report from the session of the Parliamentary Defence Committee; the numbers were provided by Defence Minister Klich, 10 January 2008. One must bear in mind, however, that many soldiers went to the mission more than one time.

14. *JDW,* 7 November 2007, 34.

15. Podbielski, *Transform or Modernize.*

16. Ibid.

17. Compare J. Szmajdzinski, "Zawodowcy" ["Professionals"], *Rzeczpospolita*, 4 October 2004.

18. *National Security Strategy,* Warsaw, 2007, paragraph 99.

19. That is, the navy could be reduced from 13,000 personnel to 10,000; the air force from 30,000 to 20,000; and the army from 80,000 to 70,000. *Raport WTO*, March 2006, 6.

20. *Wizje sił Zbrojnych*—materiały z konferencji, Biblioteka Bezpieczeństwa Narodowego, BBN, 2006, s. 62 (Visions of Armed Forces—conference proceedings published by the National Security Bureau), at http://www.bbn.pl.

21. *Rzeczpospolita*, 6 April 2007.

22. *Wizje sił Zbrojnych*, 27.

23. *National Security Strategy*, Warsaw, 2007, paragraph 98.

24. "Nowa jakość sił zbrojnych," 8 November 2007, at http://www.bbn.gov.pl/portal/pl/2/1116/Nowa_jakosc_Sil_Zbrojnych_RP.html.

25. Ibid.

26. Ibid.

27. Ibid.

28. Urbaniak, "Nowa Odsłona."

29. Ibid.

30. *Wizje sił Zbrojnych*, 27.

31. *National Security Strategy,* Warsaw, 2007.

32. Ibid., paragraph 95.

33. *Wizje sił Zbrojnych*, Brig. Gen. Andrzej Lelewski, chief of the Operational Planning Department of the General Staff.

34. The document was published by a daily "Dziennik," at http://www.dziennik.pl/wydarzenia/291363.html?service=print. Translation O.O.

35. D. Lowe, S. Ng, "Effects-Based Operations: Language, Meaning and the Effects-Based Approach," paper presented at the 2004 Command and Control Research and Technology Symposium, "The Power of Information Age Concepts and Technologies," Defence Science and Technology Organisation, Department of Defence, Canberra at http://www.dodccrp.org/events/2004_CCRTS/CD/papers/207.pdf.

36. P. Davis, *Effects-Based Operations: A Grand Challenge for the Analytical Community* (Santa Monica, CA: RAND, 2001).

37. *Nowa Jakość.*

38. See http://www.jfcom.mil/about/experiments/mne5.html.

39. *Nowa Jakość.*

40. Compare http://www.jftc.nato.int/.

41. *Wizje sił Zbrojnych.*

42. Ibid., Brig. Gen. Andrzej Lelewski.

43. *Polska Zbrojna* 49 (2 December 2007).

44. A bulletin of the National Defence Committee of the Parliament, No. 1741/V kad, 15 March 2007, 11.

45. An official working with the army put it this way: "[We] may have a corps of 'professionals' equipped with Kalashnikovs and entrenching tools." Confidential interview, Warsaw, November 2007.

9. Terriff and Osinga: Conclusion

1. Terry Terriff, interview with senior ACT official, April 2007.

2. ACT initially looked first to the US as the model for transformation but shortly recognized that American concepts were a poor fit, if not entirely unsuited, as conceptual models for European military transformation because of the real differences between American and European national interests, strategic and military cultures, resources, and domestic political environments. Terriff, interview with ACT officials, April 2004.

3. At the time of writing, both the US FCS and the British FRES programs have been canceled.

4. See, for example, Judy Dempsey, "Poland Seeks Missiles Regardless of Shield," *New York Times*, 29 May 2009, at http://www.nytimes.com/2009/05/29/world/europe/29iht-shield.html?ref=world&pagewanted=print.

5. Such a rethinking might have involved, for example, moving toward more medium-weight capabilities that were more readily transportable, up to and including the type of complete reconceptualization that the US, British, and French armies did in developing their FCS, FRES, and Scorpion networked, medium-weight armor programs.

6. Gen. James N. Mattis, "USJFCOM Commander's Guidance for Effects-based Operations," *Parameters*, Autumn 2008, 18, at http://www.carlisle.army.mil/usawc/parameters/08autumn/mattis.pdf.

7. See, for example, US Secretary of Defense Robert M. Gates, speech at Maxwell-Gunter Air Force Base, Montgomery, AL, 15 April 2009, at http://www.defenselink.mil/speeches/speech.aspx?speechid=1344.

8. See, for example, Thomas Harding, "Army Must Change or Risk Failure, Warns Future Chief," *telegraph.co.uk*, 25 June 2009, at http://www.telegraph.co.uk/news/newstopics/onthefrontline/5626220/Army-must-change-or-risk-failure-warns-future-chief.html.

9. On this problem, see Col. Ian Hope (Princess Patricia's Canadian Light Infantry), *Unity of Command in Afghanistan: A Forsaken Principle of War*, Strategic Studies Institute, "Carlisle Papers in Security Strategy," November 2008.

Index

ACT, *see* Allied Command
Transformation
Afghanistan: British forces in, 27,
49–50, 54–55, 193; communications
technology, 40, 43; Dutch forces in,
123, 124, 139–40, 193; French forces in,
66, 81, 191; German forces in, 89, 95,
96; lessons from operations in, 3–4,
16, 40, 43, 54, 139–40, 178, 193; NATO
forces in, 74, 75, 207; Polish forces in,
172, 179, 181; Provincial Reconstruction
Teams, 123, 139, 140; Spanish forces in,
154; unmanned aerial vehicles, 43; US
war in, 3, 27–28
Aircraft: Eurofighter, 52; F-117 stealth
fighters, 19, 23; navigation systems, 22;
reconnaissance missions, 22; targeting
systems, 24–25; unmanned aerial
vehicles, 19, 22, 27, 28, 43
Alliance theory, 11–12, 111, 206–8. *See also*
NATO
Allied Command Transformation (ACT),
NATO: challenges, 192; commanders,
31; Concept Development and
Experimentation, 99; Dutch military
and, 135, 136; effects-based approach to
operations, 32–33, 182; establishment,
1, 14, 55, 187; influence on allies, 14–15,
55, 55 (fig.), 108–9, 123, 135, 140–41, 143,
152, 165, 171–72, 191, 204–5; mandate,

187; Multinational Experimentations,
72, 75, 96–97, 182, 193; tasks, 207;
transformation plans, 30–34, 192; US
Transformation program and, 249n2
Aznar, José Maria, 147, 150, 166

Balkans: air operations, 130; Bosnia, 145,
154, 179; lessons from experiences in,
26, 129, 141; peacekeeping operations,
129; Polish forces in, 179. *See also*
Kosovo
Belka, Marek, 176
Bentegeat, Henri, 78, 79
Bosnia, NATO missions in, 145, 154, 179
Boyd, John, 19
Breemen, Henk van den, 118
Britain: civil-military relations, 35, 50,
197; Conservative governments, 36;
Defence White Papers, 35, 37, 43,
51; defense budgets, 36–37, 57–58;
foreign policy, 51; House of Commons
Defence Committee, 51, 54; Ministry
of Defence, 37, 39–40, 53, 57–58; New
Labour government, 36–37, 51, 197;
relations with United States, 11, 36, 38,
205; Strategic Defence Reviews, 36–37,
197
British Army: Cold War legacy forces, 36;
expeditionary capabilities, 53–55, 58;
Future Land Operational Concept,

Lightning Source UK Ltd.
Milton Keynes UK
UKHW012108030222
398167UK00001B/59

9 780804 763783